March 15–16, 2012
Zürich, Switzerland

I0047734

**Association for
Computing Machinery**

*Advancing Computing as a Science & Profession*

# MobiOpp'12

Proceedings of the 3rd ACM International Workshop on
**Mobile Opportunistic Networks**

*Sponsored by:*
***ACM SIGMOBILE***

*Supported by:*
***Disney Research, ETH, University of Cambridge, and IFIP***

**Association for
Computing Machinery**

*Advancing Computing as a Science & Profession*

**The Association for Computing Machinery**
2 Penn Plaza, Suite 701
New York, New York 10121-0701

**Notice to Past Authors of ACM-Published Articles**
ACM intends to create a complete electronic archive of all articles and/or other material previously published by ACM. If you have written a work that has been previously published by ACM in any journal or conference proceedings prior to 1978, or any SIG Newsletter at any time, and you do NOT want this work to appear in the ACM Digital Library, please inform permissions@acm.org, stating the title of the work, the author(s), and where and when published.

**ISBN:** 978-1-4503-1208-0 (Digital)

**ISBN:** 978-1-4503-1722-1 (Print)

Additional copies may be ordered prepaid from:

**ACM Order Department**
PO Box 30777
New York, NY 10087-0777, USA

Phone: 1-800-342-6626 (USA and Canada)
+1-212-626-0500 (Global)
Fax: +1-212-944-1318
E-mail: acmhelp@acm.org
Hours of Operation: 8:30 am – 4:30 pm ET

Printed in the USA

# ACM MOBIOPP 2012 Chairs' Welcome

A warm welcome to the *Third ACM International Workshop on Mobile Opportunistic Networks – MobiOpp'12*. We are very excited to introduce you to the technical program of our workshop, which includes 9 papers, 5 demos and 4 posters representing high-quality research conducted over the broad spectrum of topics related to opportunistic networking.

This year's MobiOpp received 44 submissions, coming from all continents (Brazil, India, Japan, Korea, Europe and USA). The task of providing a technical program was challenging and time-consuming, but ultimately a very rewarding one for the Technical Program Committee (TPC). Each paper was assigned to, at least, three TPC members and received reviews from either TPC members or external reviewers carefully selected by TPC members. The acceptance rate was approximately 36%. Authors of accepted papers are distributed across different regions as follows: Europe, Middle East, and Africa (67.8%), Asia/Pacific Area (16.9%), United States (13.9%), and Latin America (1.7%).

The conference program has been structured into three technical sessions, namely Social Forwarding, Analysis of Opportunistic Protocols, and Opportunistic Network Applications. Moreover, we organized two invited keynote sessions and a poster and demonstration session.

In closing, we would like to thank the TPC members and the external reviewers for their invaluable work during the review process. Special thanks go to Chiara Boldrini and Daniele Puccinelli, the Poster and Demo Co-chairs for handling the review process in a timely manner. We would also like to thank the Publicity Chairs Valerio Arnaboldi and Aruna Balasubramanian, for their long efforts in advertising the conference across the globe and attracting the large number of submissions that allowed us putting together this excellent technical program. Furthermore, we would like to express our gratitude to the Web Chair Ilias Leontiadis and the Local Organizer Andreea Picu. Also, special thanks go to Adrienne Griscti from ACM for organizing the paperwork and Lisa Tolles for taking care of processing the papers in a timely manner. Finally, we thank ETH Zürich and the University of Cambridge for hosting the workshop and website, our sponsor, ACM Sigmobile, and our generous corporate supporter, Disney Research.

We sincerely hope you will enjoy the MobiOpp 2012 program!

**Bernhard Plattner**
*MobiOpp'12 General Chair*
*ETH Zürich*

**Cecilia Mascolo and Franck Legendre**
*MobiOpp'12 Program Co-chairs*
*University of Cambridge and ETH Zürich*

# Table of Contents

**Session 4: Demos and Posters**
Session Chair: Chiara Boldrini *(IIT-CNR)*

# ACM MobiOpp 2012 Workshop Organization

**General Chair:** Bernhard Plattner *(ETH Zürich)*

**Technical Program Chairs:** Cecilia Mascolo *(University of Cambridge)*
Franck Legendre *(ETH Zürich)*

**Demo and Poster Chairs:** Chiara Boldrini *(IIT-CNR)*
Daniele Puccinelli *(University of Applied Science – SUPSI)*

**Local Arrangements Chair:** Andreea Picu *(ETH Zürich)*

**Web and Publicity Chairs:** Ilias Leontiadis *(University of Cambridge)*
Valerio Arnaboldi *(IIT-CNR)*
Aruna Balasubramanian *(University of Washington)*

**Steering Committee Chair:** Marco Conti *(IIT-CNR)*

**Steering Committee:** Jon Crowcroft *(University of Cambridge)*
Serge Fdida *(UPMC Sorbonne Université)*
Mario Gerla *(UCLA)*
Sergio Palazzo *(University of Catania)*
Mani B. Srivastava *(UCLA)*

**Program Committee:** Mostafa Ammar *(Georgia Institute of Technology)*
Aruna Balasubramanian *(University of Washington)*
Nilanjan Banerjee *(University of Arkansas, Fayetteville)*
Tim Brown *(University of Colorado Boulder)*
Tracy Camp *(Colorado School of Mines)*
Augustin Chaintreau *(Columbia University)*
Mooi Choo Chuah *(Lehigh University)*
Vania Conan *(Thales Communications)*
Jon Crowcroft *(University of Cambridge)*
Jakob Eriksson *(University of Illinois at Chicago)*
Serge Fdida *(UPMC Sorbonne Université)*
Roy Friedman *(Technion)*
Laura Galluccio *(University of Catania)*
Mario Gerla *(UCLA)*
Silvia Giordano *(University of Applied Science – SUPSI)*
Theus Hossmann *(University of Cambridge)*
Pan Hui *(Deutsche Telekom Laboratories)*
Stratis Ioannidis *(Technicolor)*
Merkourios Karaliopoulos *(National & Kapodistrian University of Athens)*

# MobiOpp 2012 Sponsor & Supporters

Sponsor:

Supporters:

ETH
Eidgenössische Technische Hochschule Zürich
Swiss Federal Institute of Technology Zurich

# Random Graph Models for Wireless and Social Networks

## [Keynote Talk Abstract]

Matthias Grossglauser
School of Computer and Communication Sciences
EPFL
Lausanne, Switzerland
matthias.grossglauser@epfl.ch

## ABSTRACT

Operating large-scale social applications over opportunistic wireless networks entails many fascinating engineering challenges. We strive for robust and efficient algorithms for specific problems like opportunistic forwarding, routing, or publish-subscribe, and we want to ascertain global properties like security, privacy, fairness, and high performance. One particular set of challenges concerns the scalability of this whole endeavor: is it fundamentally possible for such applications and underlying methods to scale up to large networks, without jeopardizing desirable system properties?

We discuss recent progress in some of the key problems in this area at three conceptual layers: opportunistic forwarding, routing under mobility, and social network privacy.

Opportunistic forwarding exploits the random broadcast nature of the wireless channel and the availability of multiple "good" routes towards a destination. This approach can deliver a message to its destination at a potentially lower expected cost than over a single shortest path. We introduce a forwarding algorithm and associated "anypath metric" that is optimal, building on the observation that no single-path metric can achieve optimality in general.

In routing under mobility, the key challenge is to keep track of the changing network topology, so that efficient routes can be computed at any time between any pair of nodes. We ask whether there exist low-overhead schemes that produce low-stretch routes, even in large networks where all the nodes are mobile. We present a scheme that maintains a hierarchical structure within which constant-stretch routes can be efficiently computed between every pair of nodes. The scheme rebuilds each level of the hierarchy periodically, at a rate that decreases exponentially with the level of the hierarchy, and achieves constant stretch under a mild smoothness condition on the mobility process.

Finally, we address the problem of the privacy of an anonymized social network. The specific challenge is the sharing or public release of anonymized network data without accidentally leaking personally identifiable information (PII). Unfortunately, it is often difficult to ascertain that sophisticated statistical techniques, potentially employing additional external data sources, are unable to break anonymity. We show an asymptotic condition, based on a random graph model, under which a computationally powerful adversary would be able to re-identify the anonymized node identities. This has important implications for privacy policies in social network structures.

What binds the above problem formulations and results together is our reliance on stochastic network models that retain only the salient features of each problem. These abstractions allow us to make precise statements about scalability to very large systems. We hope that these results complement and inform more focused work on methods, protocols, and applications for mobile, opportunistic, and social networks.

## Categories and Subject Descriptors

C.2.m [**Computer-Communication Networks**]: Miscellaneous; G.2.2 [**Graph Theory**]: Network problems

## General Terms

Algorithms, Performance, Theory

## Keywords

Wireless networks, social networks, scalability, mobility, routing, opportunistic forwarding, graph embedding, network privacy

# 404 Not Found? – A Quest For DTN Applications

Jörg Ott
Aalto University
Department of Communications and Networking
Otakaari 5A, 02150 Espoo, Finland
jo@comnet.tkk.fi

## Categories and Subject Descriptors

C.2.2 [**Computer Communication Networks**]: Network Protocols; C.2.4 [**Distributed Systems**]: Distributed Applications

## Keywords

Delay-tolerant Networking; DTN; Routing; Mobility, Applications

## 1. EXTENDED ABSTRACT

Delay-tolerant Networking (DTN) [3] has moved a long way from its—for many probably somewhat elite—origin of a technology for an Interplanetary Internet in the late 1990s to an established research discipline. One major contributor to this trend was the observation that quite a few terrestrial networks exhibit delay-tolerant properties, albeit of different nature: from sparse mobile ad-hoc to sensor networks to mobile Internet access, we find delay tolerance as an important element to describe communication behavior and to design protocols suitable for operation in the respective challenged networking environment. And even fixed infrastructure networks may benefit from delay-tolerant approaches to data transmission [7]—as did UUNet decades ago, when forwarding mails and news in a multi-hop store-and-forward fashion enabled communication involving machines that were not "always on" in the first place.

Delay-tolerant networking has contributed to our understanding of networking at least in a twofold manner:

- It has helped extending the reach of communication into areas previously beyond the grasp of generic networking architectures (as opposed to closed application-specific solutions that may have existed in some areas) [3].

- We have learned reconsidering protocol design practices to be able to build systems for DTNs,[1] which are also applicable to the Internet at large: from robustness in the presence of disruptions to separating functions for application data units from their delivery protocols (e.g., securing the objects, not the transport) to storing (i.e., caching) meaningful (and identifiable) information units [11].

The former had a substantial impact on mobile ad-hoc networking research for which mostly connected networks (with high node density) had to be assumed, a requirement that could be relaxed with delay tolerance so that MANETs got much closer to reality.

---

[1] Various issues have arisen independently in other contexts.

*MobiOpp'12*, March 15–16, 2012, Zürich, Switzerland.
ACM 978-1-4503-1208-0/12/03.

While also applicable to sparse vehicular networks or wireless sensor networks, enabling networking between mobile devices, also dubbed *opportunistic* or *pocket-switched networking* [5], has been the primary driver for research in a number of fields. These include studying contact patterns between mobile devices carried by humans, understanding human mobility (as far as relevant for device-to-device communication) and developing models that can reproduce the observed contact patterns, and correlating the contacts and social networks to allow prediction of future interactions.

These and others feed into system and protocol design for mostly infrastructure-less communication between mobile nodes and their respective evaluation. Dozens of routing protocols were developed and diverse communication paradigms beyond unicast adapted to mobile DTNs, including multicast, publish/subscribe, and broadcast, with many different flavors being developed. Recently, plain opportunistic communication got another layer on top: service discovery and having tasks executed by individual or groups of other nodes to later collect the results: *opportunistic computing* [2] or *crowd computing* [10], in analogy to cloud computing.

So, here we are with all these models and protocols and systems. But what to do with them? Once in a while it feels that, with delay-tolerant networking, we are heading for the same trap that the MANET community has fallen into before: designing (routing) protocols without any clear applications in mind that might use them or defining more or less artificial problems with little bearing on reality. Disaster management, crisis scenarios, and communication in remote areas are prime examples for application scenarios to be found in grant applications. Yes, these scenarios do matter, but they cannot be all there is. We should do better—and we can!

However, we must always remember what or whom we are designing for and which requirements arise from this target. For example, a lot of work has explored human-to-human communication or Internet access in opportunistic environments: for web page retrieval, web and vicinity search, twitter, and messaging, among others. But very little has been done to understand how such applications would need to be designed to become actually usable and useful. We often tend to be happy about (or blinded by?) performance figures that show that something works *in principle* without questioning too much the implications *in practice*. The latter particularly includes appreciating that the delay tolerance of the user and her interaction with the system (and thus the user interface).

This touches a key point concerning opportunistic networks: the lack of predictability. If users want something, they want it now! At the very least, they want to know when to expect a response. "404 Not Found" or a browser's note "Server not found" are clear, whereas an hourglass or a similar indication to wait leaves the user in the dark about her chances for success. Opportunistic networking environments in which message delivery latency can be anything between one second and several days—or infinity since odds are that a message is not delivered at all—deny a user this very predictability.[2] Users calibrate their expectations anyway when directly connected to the Internet, when search results or the initial bits of web page contents are expected to appear within a second—and our work culture has evolved to expect even asynchronous means for information exchange such as email to work more or less instantaneously.

All this makes building compelling DTN applications a challenging task, given such demanding competition especially in places with pretty good wireless or cellular coverage and flat rates for mobile data at least for local subscribers. We have basically two options: 1) Reconsider the way we think about application interaction and communication paradigms that can satisfy mainstream user expectations. 2) Find those *niches* where the mainstream does not matter. The aforementioned disaster, crisis, and rural communication applications fall into this latter category.

For the mainstream, we face the challenge of designing applications that either cannot be built using omnipresent wireless infrastructure and backend services in the cloud or try to minimize leaking information to such services. Exchanging large volumes of information at high data rates or in privacy, implicitly provided by physical proximity to some extent, could thus be important drivers.

One obvious use is **content sharing**. DTN content distribution has been pioneered by the *PodNet* system [8] that offers users to share content in a peer-to-peer fashion according to their interests. *Floating Content* [12] restricts content dissemination to a predefined geographic area for each content item and thus exploits locality, supports spatial and temporal re-use of node buffers, and implicitly limits resource consumption. *SCAMPImusic* allows experiencing the music tastes of a user's immediate surroundings by sharing music contents in a volatile way.[3] From a usability perspective it is important that these applications all work in the background, collecting content according to a user's preferences: when they catch the user's attention, they only present what they already got, thus preserving the instant interaction a user would expect. Moreover, as there is no way for a user to know which content to expect in each case, no expectations (e.g., completeness of the available music) can be disappointed.

Content sharing between mobiles may **complement infrastructure** usage. Devices may perform opportunistic caching and ask their neighbors about locally available content before downloading via the infrastructure. This may *offload* data from the infrastructure, but also assist in preserving privacy, e.g., when sharing map tiles of the environment so that server cannot track users [1]. Such applications would operate in the background invisible to the user.

Niche applications have contributed substantially to DTN development, especially before mobile phones were close enough to the capabilities required for ad-hoc interactions between them. In the past, quite some focus was on **sensing**, e.g., for environmental conditions [9] or to study animal behavior [6], where sensor data were replicated between nodes and collected at dedicated places, using animal, vehicle, or human mobility for data carriage. Many further applications to extend the reach of networking for different areas (underwater, mountains, aerial surveillance) continue to emerge.

Another niche application area has received rather limited (academic) attention so far: various **industrial environments** may offer quite similar communication challenges as mountainous or ru-

ral areas, such as lack of (deployable) infrastructure, sparse node density, and too limited communication range. They would benefit from delay-tolerant networking to improve (if not enable) communications. One example is networking people, equipment, and a control room in underground mines [4], where rock blocks wireless communication effectively. Mines have usually very limited facilities for data communications and, if available, cover only a fraction of the total mine. Yet orders need to be sent to and status information collected from potentially hundreds of machines so that progress monitoring and planning can proceed efficiently. Leveraging vehicles and workers equipped with mobile devices as message carriers can improve information availability substantially. Such constrained environments, where DTN can make a difference in practice, can likely be found elsewhere.

While niche applications may benefit from dedicated hardware and closed deployments, which simplifies configuration and operations, mainstream applications face an open and heterogeneous environment. Especially with content sharing, making participation legally safe, technically robust (e.g., against malware), and privacy preserving are challenges yet to be addressed.

One common observation across all applications is that they are usually applied in a specific context and those observations and requirements don't find their way into other researchers' evaluations too often. So, we need to break out of these research silos to share (and use!) insights about meaningful scenarios, applications, and parameters more broadly: too often, we find evaluations for routing protocols that use traffic with little relation to applications or a rather limited set of mobility models. And evaluations are often quite optimistic about lower layer characteristics from detection times, to pairing success rates, to transmission rate.

Finally, it is difficult to obtain real usage patterns for DTN applications without developing, deploying and measuring them. We need to "eat (more of) our own dog food" to gain a better understanding of how those applications and their underpinnings work for real. If we don't, who would?

## 2. REFERENCES

[1] AMINI, S., LINDQVIST, J., HONG, J., LIN, J., TOCH, E., AND SADEH, N. Caché: Caching location-enhanced content to improve user privacy. In *Proc. of ACM MobiSys* (2011).

[2] CONTI, M., GIORDANO, S., MAY, M., AND PASSARELLA, A. From Opportunistic Communication to Opportunistic Computing. *IEEE Communications Magazine 48*, 9 (September 2010), 126–139.

[3] FALL, K. A Delay-Tolerant Network Architecture for Challenged Internets. In *Proceedings of ACM SIGCOMM* (August 2003).

[4] GINZBOORG, P., KÄRKKÄINEN, T., RUOTSALAINEN, A., ANDERSSON, M., AND OTT, J. DTN communication in a Mine. In *Proc. of ExtremeCom* (September 2010).

[5] HUI, P., CHAINTREAU, A., SCOTT, J., GASS, R., CROWCROFT, J., AND DIOT, C. Pocket Switched Networks and Human Mobility in Conference Environments. In *ACM WDTN Workshop* (2005).

[6] JUANG, P., OKI, H., WANG, Y., MARTONOSI, M., PEH, L. S., AND RUBENSTEIN, D. Energy-efficient computing for wildlife tracking: Design tradeoffs and early experiences with zebranet. In *Proc. of ASPLOS-X* (October 2002).

[7] LAOUTARIS, N., AND RODRIGUEZ, P. Good Things Come to Those Who (Can) Wait. In *Proc. of ACM HotNets VII* (2008).

[8] LENDERS, V., MAY, M., KARLSSON, G., AND WACHA, C. Wireless ad hoc podcasting. *ACM/SIGMOBILE Mobile Comp. and Comm. Rev. 12*, 1 (2008).

[9] MCDONALD, P., GERAGHTY, D., HUMPHREYS, I., FARRELL, S., AND CAHILL, V. Sensor Network with Delay Tolerance (SeNDT). In *Proc. of IEEE ICCCN* (2007).

[10] MURRAY, D. E., YONEKI, E., CROWCROFT, J., AND HAND, S. The Case for Crowd Computing. In *Proc. of ACM MobiHeld* (2010).

[11] OTT, J. Delay Tolerance and the Future Internet. In *Proc. of IEEE WPMC* (2008).

[12] OTT, J., HYYTIÄ, E., LASSILA, P., VAEGS, T., AND KANGASHARJU, J. Floating Content: Information Sharing in Urban Areas. In *Proc. of IEEE Percom* (2011).

---

[2]This also applies to the aforementioned applications: e.g., people in emergency situations want to know if somebody heard them and need a "voice" to tell them that help is on the way.

[3]Idea and prototype developed by Teemu Kärkkäinen in the EC FP7 project SCAMPI (grant agreement 258414).

# On Context Awareness and Social Distance in Human Mobility Traces

Anna Förster, Kamini Garg, Hoang Anh Nguyen, Silvia Giordano
NetLab, ISIN-DTI, SUPSI, Manno, Switzerland
[anna.foerster,kamini.garg,hoang.nguyen,silvia.giordano]@supsi.ch

## ABSTRACT

Opportunistic networks consist of mobile devices, carried by people in their everyday lives. They organize autonomously to exchange data with direct neighbors without the use of any infrastructural services. Since the devices are carried by humans, one of the main challenges to consider in opportunistic networks is the human mobility behavior. However, little work exists on how the social behavior of people drives their mobility behavior and how this context information can be systematically leveraged for opportunistic networking applications.

This paper tackles this problem by providing both experimental and theoretical analysis of human mobility context information. We present a novel real world experiment with sensor nodes carried by people to demonstrate and study the effect of context on people mobility. Furthermore, we define a novel metric of *social distance* to put this new evidence on solid mathematical foundation. Thus, our work puts a basis to systematically leveraging context information for opportunistic networking applications and services. Additionally, our experimental data traces enable testing and evaluation of such novel services in a real world scenario.

## Categories and Subject Descriptors

C.2.1 [**Computer-Communication Networks**]: Network architecture and design- *design, protocols, routing, scheduling, security, wireless sensor networks*

## General Terms

Algorithms, Experimentation, Performance, Theory

## Keywords

opportunistic networks, sensor networks, mobility traces, context-aware traces, evaluation, simulation

## 1. INTRODUCTION

Opportunistic networks have a great potential to leverage everyday mobile devices and to enable very large and infrastructure independent networks. Such networks have the potential to provide communication to remote and disconnected regions and to enable novel applications and services without depending on existing infrastructures.

Some state of the art work in opportunistic networking, such as our own routing protocols [13], assume the availability of some kind of context information for the involved users. This information ranges between frequented locations, such as job or home location, to interests and hobbies. However, this context information was not sufficiently explored in the literature, neither in theoretical nor in experimental way.

This paper addresses this gap and presents both: an experiment for gathering human mobility traces with rich context information and a mathematical definition of *social distance* between people that is then validated with our traces. This work builds the foundation for further context-oriented opportunistic networking research by providing experimental evidence for its relevance, as well as a theoretical basis.

In this paper, we first elaborate on existing work on human mobility traces and especially on the availability of context information in them. Then, we present our experimental approach and setup in Section 3, where we gather context-rich mobility traces for over 3 weeks with 39 people. We dive into the obtained results, their meaning and statistical evaluation in Section 4. Then, we generalize our findings by defining a new metric called *social distance* in Section 5. Section 6 ties back to our experimental results and presents evidence for the significance of the newly defined metric. Section 7 finally sketches our future directions and concludes the paper.

## 2. BACKGROUND AND MOTIVATION

While working in the area of opportunistic networking and developing routing protocols for opportunistic networks, our intuition has been that users with similar interests or social behavior have greater probability to meet each other and to exchange data. Thus, we tried to leverage this social profile knowledge in our protocols, e.g. in Propicman [13]. However, a major challenge is to evaluate such routing approaches. Implementing it straight away for real devices is too risky and time-consuming, even though it will be the perfect evaluation environment, with existing social behavior of the participants. On the other side, simulation offers the possibility for structured evaluation and debugging at

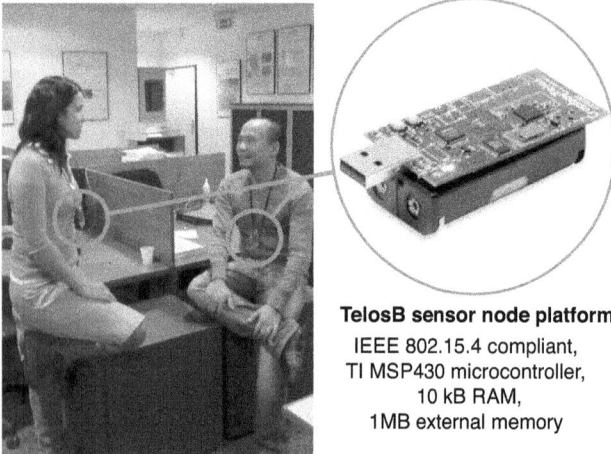

**TelosB sensor node platform**
IEEE 802.15.4 compliant,
TI MSP430 microcontroller,
10 kB RAM,
1MB external memory

Figure 1: General experimental setup: people carry sensor nodes around their necks, which broadcast beacons every 10 msec with transmission range approx. 2-3 meters. On the right the used sensor node platform TelosB from Memsic.

low cost. However, simulation models used to mimic the behavior of device carriers are mostly location-based.

Generally speaking, there are two types of mobility models that can be used: traces and synthetic models [1]. There exist also a third alternative, where synthetic models are build upon evaluating distributions of real traces. Several experiments have been performed to obtain such real traces. In general, a set of mobile nodes (humans [5], animals [6], vehicles [19]) is equipped with mobile devices capable of measuring either directly contacts to other devices or their own current location. Most of these real traces are freely available in the CRAWDAD database (http://crawdad.cs.dartmouth.edu/index.php).

Synthetic models can be based on social network theory (such as the Community based mobility model – CMM [12]) or on contact and trajectory models (such as SLAW [11]). CMM allows collections of hosts to be grouped together in a way that is based on social relationships among the individuals. This grouping is only then mapped to a topographical space, with topography biased by the strength of social ties. The movements of the hosts are also driven by the social relationships among them. The model also allows for the definition of different types of relationships during a certain period of time. Home-cell CMM (HCMM [2]) extends CMM by adding to the social relationships also attraction of physical places. This idea was already used in [9], where each user is assigned to a community that represents a frequently visited place, and moves around with transitions governed by a two-state Markov chain.

In the Working Day Movement Model [7], the social aspects are represented by routines (go to work in the morning, spend their day at work, and commute back to their homes at evenings). The model combines several elements to reproduce the social behaviour. In the Agenda Driven Mobility Model [18] the authors include the social activities in the form of agenda (when, where and what). They are able to capture most of the human mobility characteristics in the observed scenario, but the model is mixing all those characteristics as global driving force, and the resulting solution

is both very complex as well as difficult to apply in other scenarios.

The contact and trajectory models build upon the characterization of the contact time, the inter-contact time and the trajectories derived by real traces. The traces collected within Haggle experiments [15] allowed to derive power-law inter-contact time distributions with cut off [3] [10], the levy-walk nature of human mobility consisting of sets of small moves followed by long jumps [14], the scattering distribution of visited places [11] and other interesting properties that were used to define synthetic mobility models that adhere to one or more of those distributions. For example, SLAW [11] mobility model is characterized by a truncated power law distribution for the distance selection of consecutive movements and is based on the existence of some preferred locations, which are more popular than others to a certain user. SWIM [17] starts from the observation in [8] where a high degree of temporal and spatial regularity is demonstrated, which is in contrast with the random trajectories of previous models. SWIM captures this social behavior and selects the destination points based on their popularity among all nodes and their distance from the home point.

However, to the best of our knowledge, there does not exist a simulation model or real wireless trace, which incorporates also social behavior information, such as food preferences, spoken languages, sport interests. Such information cannot be directly associated with locations or geographic position in general and needs to be captured separately as a driving force to human mobility patterns and especially inter-human contacts.

## 3. USING SENSOR NODES TO CAPTURE CONTEXT-AWARE MOBILITY TRACES

The main motivation of our experiment was to gather real human traces with context information. To the best of our knowledge, such traces are not readily available, as typical mobility traces include only contact or geographic information. However, our opportunistic routing protocols make use of social context information, such as interests or preferences and we needed to prove that such social information is indeed relevant to the mobility trace of individual people.

### 3.1 Experiment description

A total of 39 people from three different institutes, located in two different buildings participated in our social-context mobility experiment. We chose to use wireless sensor nodes of the type TelosB [1] rather than smart phones for battery lifetime and cost efficiency. Each person receives a sensor node to be carried around the neck (see also Figure 1) with batteries, which hold up to 2 days, depending on traffic.

The sensor nodes are configured as to have a transmission radius of around 5 meters, thus nodes of people in the same or adjacent rooms can communicate with each other. A neighbor discovery protocol sends beacons via one-hop broadcast every 10 milliseconds. If a node receives a beacon from some other node, it answers with a reply-beacon. All beacon receptions are logged into the local memory of the sensor nodes and downloaded regularly to several base stations, scattered around in the buildings. Note that logged communication can be both unidirectional and bi-directional

---

[1] www.memsic.com

| Property | Possible answers (#answers) |
|---|---|
| **Location-based properties** | |
| Institute | WG1(6), WG2(16), WG3(17) |
| Building | B1(10), B2(28) |
| Office | F1(5), F2(2), F3(3), F4(3), F5(3), F6(2), F7(2), F8(3), F9(2), F10(4) |
| LunchPlaces | P1(7), P2(10), P3(13), P4(9), P5(9), P6(15) |
| **Social-based properties** | |
| Job position | J1(6), J2(22), J3(4) |
| Spoken languages | L1(29), L2(35), L3(2), L4(12), L5(11), L6(2), L7(2) |
| Vehicle | V1(24), V2(12), V3(6) |
| Smoking | S1(5), S2(33) |
| BreakDrinks | B1(24), B2(11), B3(11) |
| Music Preferences | M1(2), M2(9), M3(10), M4(10), M5(4) |
| Sports interests | SP1(3), SP2(3), SP3(4), SP4(2), SP5(3), SP6(2), SP7(8), SP8(5), SP9(2), SP10(3) |
| Joint projects | R1(2), R2(3), R3(3), R4(4), R5(2), R6(7), R7(3), R8(2), R9(2), R10(3), R11(3), R12(5), R13(2), R14(2), R15(3), R16(4) |

Table 1: All context profile information for all participants in the experiment. The properties are divided into location-based and social-based and the number of answers to each of the possible values is provided in brackets.

| Timestamp (msec) | Sender (ID) | Receiver (ID) |
|---|---|---|
| 543900 | 2 | 15 |
| ... | ... | ... |

Table 2: The log format of our contact traces. The timestamp is given in milliseconds from experimental start, the sender and receiver with their identifier numbers (ID). IDs are positive numbers only.

and that we capture also unidirectional links between nodes. Although the internal memory of the nodes is restricted, we have not suffered from data loss, as the download rate to the base station was high.

## 3.2 Social information

Before starting the experiment, every participant filled out a context information profile questionnaire, as depicted in Table 1. The profile consists of two types of information: location-based and social-based. The questionnaire is multiple-choice, but the choices were selected carefully with some a-priori knowledge of the participants (e.g. for spoken languages). The same Table 1 presents also the number of responses we received from the participants, e.g. 6 participants marked WG1. Real values are omitted for privacy reasons.

## 3.3 Log format and synchronization

The log format of the contact traces is depicted in Table 2. In fact, the trace contains only contact data with information when two nodes exchanged one packet. Recall that the application on the nodes broadcast beacons with transmission range of around 2-3 meters every 10 milliseconds.

Sensor nodes like the TelosB have only local time, starting from 0 every time the sensor node reboots. Thus, we needed a global synchronization mechanism. Additionally, we needed a download ability to a server throughout the experiment, as sensor nodes can only keep approx. 1 MB of data locally. We solved both issues with the implementation of base stations with a statically connected sensor node to

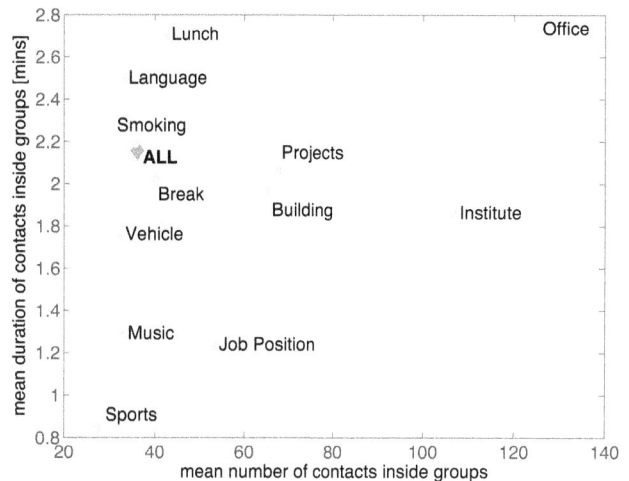

Figure 2: Comparison of mean number and mean duration of contacts for the complete trace (all, i.e. no social information, the red diamond) against individual social groups (green dots). For example, people who speak the same language have longer contacts, but less number of contacts than in the complete trace.

them. The base station provides a global timestamp to all nodes for synchronization and a download capability.

## 4. EXPERIMENTAL RESULTS

Our experiment ran with 39 people over 3 weeks in December 2010 and we gathered over 600,000 contacts between devices, see Table 3. We have evaluated and validated our traces with various statistics and metrics. Whenever two people are close enough to each other to exchange some packets, we count a *contact*. During one contact, some limited number of packets might get lost. Thus, a contact has a total *duration* and is able to exchange some *number of packets*. Note that the number of packets is not necessarily directly

| Complete trace | |
|---|---|
| **Characterics** | **Value** |
| Duration of experiment | 3 weeks |
| Number of devices | 39 |
| Total number of contact packets | 659,801 |
| Number of contacts | 17,458 |
| Mean number of contacts between two nodes | 36 |
| Mean duration of contacts | 2.2 minutes |
| Mean number of packets between two nodes | 488 packets |
| Mean inter-contact time | 128 minutes |

**Table 3: Overall experimental results for our complete contact trace.**

deducible from the duration of the contact, as some contacts are more lossy than others.

First, in Table 4 we present these metrics over the whole trace (all contacts of all people) and of isolated social communities. For example, the mean number of exchanged packets during the whole experiment between any two nodes was 488 packets. However, when considering only the contacts between any two people, who speak the same language, the mean number of packets grows by 18%. This infers that a greater probability exists for two people to meet and to exchange information, when they have similar social properties, such as speaking the same language. This kind of information is summarized in Table 4 for all social groups in our experiment and organized by location-based and social-based properties. Note that the results in terms of different metrics or different social communities do not sum up to the full trace metrics, because the same person is part of several communities.

Generally speaking, we observe that location-based communities, where people frequent the same places, increase the contact probability more significantly than social-based properties. However, some social-based properties increase the contact probability also significantly compared to the full trace, e.g. joint projects and job position. Some other properties, such as sport or music preferences, seem to have the opposite effect. This is probably due to the nature of our experiment and to its general scenario, where people work. In general, people at work tend to meet each other for fulfill their duties, e.g. joint projects. Thus, in our scenario we have some strong properties (location and social based) and some weak ones. We will later define these strong properties as *catalyst* properties, see Section 5.

In Figure 2, the differences between the contacts of different social groups are visually depicted, in terms of the duration of the contacts and their number. For example, people clearly tend to have more and longer contacts with other people in the same office.

The inter-contact time distribution was first used for opportunistic networks in [4], where the authors also showed that the inter-contact time follows a power law for some rather short periods of time. Inter-contact time is defined as the time elapsed between two consecutive contacts between two devices. The distribution of inter-contact times for our experiment is depicted in Figure 4. The slope of the resulting power law is 0.37 (see also the figure) and is very similar to what the authors in [4] have fitted for their experiments with mobile devices at two INFOCOM conferences

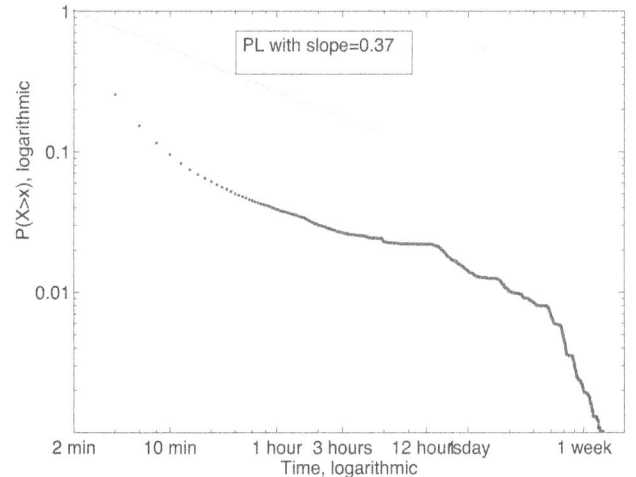

**Figure 4: Intercontact time distribution for our complete dataset. The X axis is the timeline (logarithmic scale), the Y axis shows the probability of a certain inter-contact to be more than X (logarithmic scale). The probability follows a power law until approx. 12 hours of inter-contact time and then becomes exponential.**

(traces are available at CRAWDAD [2]). This is reasonable, as a conference environment is similar in its dynamics as a campus one, where the same people are moving in the same restricted environment for most of the time. Our results also confirm the conclusion of [10] that the distribution of inter contact times follows the power law only for rather short inter-contact times, in our case up to 12 hours, and then becomes exponential.

## 5. DEFINITION OF SOCIAL DISTANCE

As we observed in our experimental data, people who share some location-based or social-based properties tend to have more contacts in general. At the same time, some properties seemed to be more potent than others. These findings are very important for future research in opportunistic and delay-tolerant networking, where people are somehow involved. Our experiments are the first to confirm this intuition and here in this section we provide clear theoretical basis to identify social communities and their shared properties, the importance of the different properties and a way to predict the number of contacts between people. In the next section we will demonstrate that our formal conclusions hold experimentally.

We consider a group of people $G = \{p_1, p_2, ..., p_n\}$. We define as *property* $P$ a quality common to all members of our group $G$. For example, $P$ can be something universally valid such as *age*, as we can associate an age to each person of any group $G$, but $P$ can also be something more specific like we had in our experiment, for example *language* or *smoking*.

Each property $P_i$ takes values in a closed set $V_i$,

$$P_i : G \to V_i \subset \mathbb{R}$$

For example, the property "living_place" could be represented by the kilometers from the geographical center of the

[2]crawdad.cs.dartmouth.edu.

| Metric | All | Location Communities | | | | Social communities | | | | | | | |
|---|---|---|---|---|---|---|---|---|---|---|---|---|---|
| | Full trace | Working Group | Building | Office | Lunch places | Job position | Languages | Vehicle | Smoking | Break | Music | Sports | Joint projects |
| Mean number of contacts | 36 | 107 | 65 | 125 | 43 | 54 | 33 | 32 | 35 | 40 | 33 | 27 | 67 |
| Difference to full trace | n/a | +196% | +81% | +247% | +19% | +48% | -7% | -9% | -1% | -6% | -7% | -23% | +86% |
| Mean duration of contacts [min] | 2.15 | 1.85 | 1.96 | 2.73 | 2.7 | 1.23 | 2.49 | 1.75 | 2.19 | 2.02 | 1.28 | 0.84 | 2.06 |
| Difference to full trace | n/a | -13% | -8% | +26% | +25% | -42% | +16% | -18% | +1% | +1% | -40% | -60% | -4% |
| Mean number of packets | 488 | 944 | 696 | 1960 | 959 | 356 | 574 | 423 | 509 | 495 | 297 | 228 | 1195 |
| Difference to full trace | n/a | +93% | +42% | +302% | +96% | -27% | +18% | -13% | +4% | -8% | -39% | -53% | +144% |

Table 4: Overall experimental results for our contact trace. Metrics are computed over the complete dataset and over individual groups only. For individual groups, the difference to the complete dataset is given rather than exact numbers. See also Figures 2–3 .

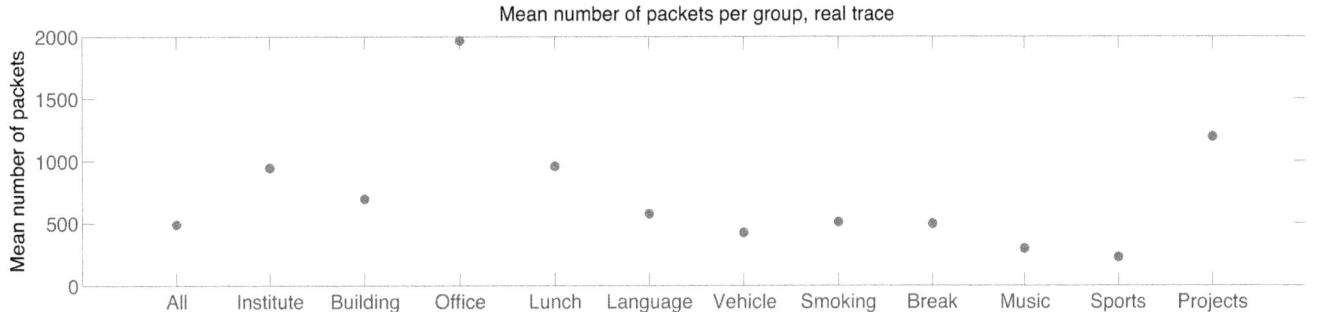

Figure 3: Mean number of individual packets, exchanged between any two nodes in the complete dataset (all, the first point) compared to people with some shared social properties.

group, or between the living place of two people, if $G$ is a group of people that meet each other. But it could be the cost of the house if the purpose of the property is to rank the people of $G$ based on an economical standpoint. In other cases, the mathematical meaning can be less straightforward, for example for spoken languages.

DEFINITION 1 (P-DISTANCE). *We define distance with respect to the property $P_i$, the function*

$$D_{P_i}(p_j, p_k) = |P_i(p_j) - P_i(p_k)|$$

P-Distance is a mathematical function, since all properties are constructed so that $V \subseteq \mathbb{R}$. In some cases, when the construction of $V$ is not artificial, P-Distance corresponds to the physical distance. For example, in our experiments, for the office property, P-Distance corresponds to the physical distance, as $V$ is really $\mathbb{R}^+$, and we compute P-Distance as the physical distance between two offices. In other cases, where $V$ is artificially mapped onto $\mathbb{R}$, this correspondence with physical distance does not hold, as for the language property. We will introduce some examples for P-Distance later in Section 6.

DEFINITION 2 (P-CLOSENESS). *We say that $p_1$ is closer*

to $p_2$ than $p_3$ with respect to the property $P_i$ if

$$D_{P_i}(p_1, p_2) < D_{P_i}(p_3, p_2) \qquad (1)$$

As we will experimentally demonstrate in Section 6, two people that are closer for given properties have also higher probability to get in contact. Furthermore, some properties tend to drive more contacts, i.e. to be more potent than others. Given that $Pr^C(p_j, p_i)$ is the probability of contact between $p_j$ and $p_i$, we define:

DEFINITION 3 (CATALYST PROPERTY). *We define a property $P_c$ in a group $G$ as catalyst, if*

$$Pr^C(p_j, p_k) > Pr^C(p_l, p_m), \quad \textbf{\textit{and}}$$

$$D_{P_c}(p_j, p_k) < D_{P_c}(p_l, p_m) \qquad with \quad p_i, p_j, p_l, p_m \in G,$$

*thus, the contact probability $Pr^C$ of $p_j, p_k$ is greater than the contact probability of $p_l, p_m$.*

For example, our property job position is catalyst, because it drives more contacts for people with the same job position than for people with different ones.

In order to further understand the relationship between closeness and contact probability we introduce a special attribute for properties, called the relevance or the weight of a specific property.

We indicate with $w_i$ the *relevance* of the property $P_i$. $w_i$ indicates the importance of the property $P_i$ within the group. For example, in a group of Internet game players, the property "age" is more relevant than the property "living_place" as the spread of electronic games is generational and not locational. $w_i \subseteq \mathbb{R}$ and thus can take also negative values.

The weight $w_i$ plays an important role for representing the social behavior with respect to a property: we consider a property is more relevant when it is also more likely to enable contacts between people. This is derived by contact or social graph theory, as discussed in [16]: the scalar weight $w_i$ is derived, at time $t$ as a function of the contact trace matrices $A(t)$ for some past window such that $w_i = f(A(t_0), A(t_0 + 1), ..., A(t))$.

Thus, by using these weights, we can scale all P-Distances and define:

DEFINITION 4 (WEIGHTED P-DISTANCE). *We define weighted distance with respect to the property $P_i$, the function*

$$D_{P_i}^{w_i}(p_1, p_2) = w_i * D_{P_i}(p_1, p_2)$$

*where the weights $w_i$ represent the relevance of the individual properties in $G$ as discussed above.*

With this new weighted definition of P-Distance we are able to define the most important distance, that is the *social distance*, which is further used to order elements within a group.

DEFINITION 5 (SOCIAL DISTANCE). *In a group $G$, the social distance between two people $p_1$ and $p_2$, with $p_1, p_2 \in G$ is defined as the sum of the weighted distances for all the properties $P_i$ of $G$ considered with their relevance within $G$.*

$$SD(p_1, p_2) = \sum_i D_{P_i}^{w_i}(p_1, p_2) \qquad (2)$$

The social distance can be seen as an extension of *physical distance* that includes the whole social context of a person within a group $G$.

DEFINITION 6 (SOCIAL CLOSENESS). *We say that $p_1$ is socially closer to $p_2$ than to $p_3$, if the social distance of $p_1$, $p_2$ is smaller than the social distance of $p_1$, $p_3$*

$$SD(p_1, p_2) < SD(p_1, p_3) \qquad (3)$$

COROLLARY 1 (SOCIAL CLOSENESS DRIVES CONTACTS). *If $p_1$ is socially-closer to $p_2$ than to $p_3$, thus the probability of contact between $p_1$ and $p_2$ is higher than the probability of contact between $p_1$ and $p_3$.*

Consequently, socially closer people build groups or communities, which we define as follows:

DEFINITION 7 (SOCIAL GROUP OR COMMUNITY). *A group $G$ is a social group or community with respect to properties $P_1, P_2, ...P_m$ if all properties are catalyst or have $w_i = 0$.*

This implies that, whenever two people in a community are closer, they have higher probability to get in contact, as validated in the next section with our experimental results.

| Property | P-Distance | Weight |
|---|---|---|
| Institute | $D = 0$, if same <br> $D = 1$, else | 1 |
| Building | $D = 0$, if same <br> $D = 1$, else | 1 |
| Office | $D = 0$, if same <br> $D = 0.5$, if same floor <br> $D = 1$, else | 1 |
| Lunch place | $D = 1 - \sum \frac{shared*}{all}$ | 1 |
| Job position | $D = 0$, if same <br> $D = 0.5$, if close in hierharchy <br> $D = 1$, else | 1 |
| Language | $D = 1 - \sum \frac{shared*}{all}$ | 1 |
| Vehicle | $D = 0$, if bus <br> $D = 1$, else | 0.2 |
| Smoking | $D = 0$, if both yes <br> $D = 1$, else | 1 |
| Break | $D = 1 - \sum \frac{shared*}{all}$ | 0.2 |
| Music | $D = 1 - \sum \frac{shared*}{all}$ | 0.2 |
| Sports | $D = 1 - \sum \frac{shared*}{all}$ | 0.2 |
| Joint projects | $D = 1 - \sum \frac{shared*}{all}$ | 1 |

*\* Shared: number of shared values, e.g. languages. All: number of total values for this pair, e.g. languages spoken by either person*

Table 5: P-distance functions for all properties in our experiment. The weights are calculated as explained in Section 5.

## 6. EXPERIMENTAL SOCIAL DISTANCE

In this section, we validate the formal definition of social closeness and Corollary 1 from the previous section by calculating the social distance between any two people in our experiment and comparing the number of their contacts against their social distance.

The main challenge is to identify the $P-distance$ function for all properties, as discussed in Definition 1. The functions we used are straightforward, summarized in Table 5. One of the mostly used functions is:

$$D = 1 - \sum \frac{shared}{all} \qquad (4)$$

Where *all* is the size of the union of all property values for two people, and *shared* is the size of the intersection of the values for both people. For example, if person $p_1$ speaks languages $\{A, B, D\}$ and $p_2$ speaks $\{A, C\}$, then $all = 4, shared = 1$. This is a simple way to compute the distance between two people, who share individual interests

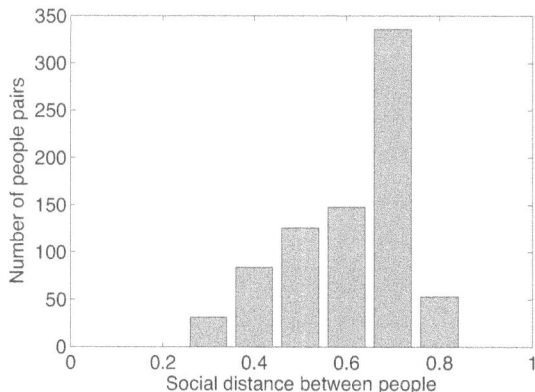

Figure 5: Histogram of computed social distances in our experimental trace between all available pairs of people. Most people seem to be not very close to each other with a social distance of 0.7.

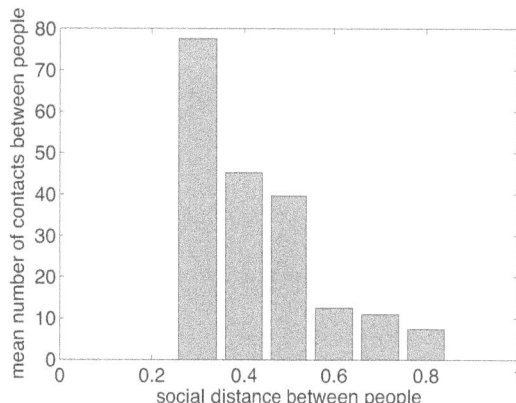

Figure 6: Dependence between social distance and mean number of contacts between people with the same social distance between each other. The closer two people are, the more contacts they have, which validates our Corollary 1.

or properties, like languages, music styles, food preferences, etc. The more values the people share, the closer they are.

Note that all $P-Distance$ functions compute the distance in the interval $[0-1]$. The weights are computed as explained in Section 4. Intuitively, they also reflect the *relevance* of the property for the contacts in a social community. For example, let us consider again Table 4. All properties, which increase the probability of contacts compared to the complete trace, have been granted a weight of 1 and all others a weight of 0.2. We applied this step function to simplify the calculations and the understanding of the weights and reflect also the catalyst properties in our experiment in a simple way.

Figure 5 presents the histogram of social distances between all pairs of people in our experiment, computed with the discussed functions and weights. The most people seem to have little in common (a social distance of 0.7), which is plausible given that we have people working on various projects and different institutes. This is confirmed by Figure 6, where we plotted the number of contacts against the social distance for all pairs of people. The closer a pair of people is, the more contacts they have. We have identified this law as Corollary 1 and Figure 6 confirms this very well.

Of course, this validation is limited to the experiment and contact data we gathered. The mathematical model, presented in Section 5 needs further validation with further experimental traces, which is part of our future work.

## 7. CONCLUSION

In this paper we tackled the problem of identifying the potential of context awareness in human mobility models for predicting contacts between people. We presented a novel experiment, where we gathered contact data of 39 people over three weeks with contact information about their interests and preferences. Furthermore, we defined a new mathematical framework to compute the *social distance* between people and to predict the number of contacts between them. We validated our framework with the traces we gathered and found out, that human context impacts significantly human mobility and especially human contacts.

Our work builds both an experimental and theoretical

basis for leveraging context information in human-centered networking scenarios, such as opportunistic networking. Furthermore, our traces are publicly available [3] and can be used by other researchers for performance evaluation and testing. In the future, we plan to further refine and validate our mathematical framework of social closeness and communities.

## Acknowledgment

We would like to thank all participants of our experiment for their effort and time. This work is partially supported by the EU Cooperating Objects Network of Excellence (CONET) and by the EU FP7 Project SCAMPI: Service platform for social Aware Mobile and Pervasive computIng (ICT grant agreement 258414).

## 8. REFERENCES

[1] Tracy Camp, Jeff Boleng, and Vanessa Davies. A survey of mobility models for ad hoc network research. *Wireless Communications and Mobile Computing (WCMC)/Special Issue on Mobile Ad hoc Networking (Research, Trends and Applications)*, 2:483–502, 2002.

[2] C.Boldrini and A. Passarella. Hcmm: Modelling spatial and temporal properties of human mobility driven by users' social relationships. *Computer Communications*, 2010.

[3] A. Chaintreau, P. Hui, J. Crowcroft, C. Diot, R. Gass, and J. Scott. Impact of human mobility on the design of opportunistic forwarding algorithms. In *Proceedings of IEEE INFOCOM*, 2006.

[4] A. Chaintreau, P. Hui, J. Crowcroft, C. Diot, R. Gass, and J. Scott. Impact of human mobility on opportunistic forwarding algorithms. *IEEE Transactions on Mobile Computing*, 6:606–620, 2007.

[5] M. Conti, S. Giordano, M. May, and A. Passarella. From opportunistic networks to opportunistic

---

[3]http://isin.dti.supsi.ch/NetLab/index.php/anna-downloads

computing. *IEEE Communication Magazine,* December 2010.

[6] Vladimir Dyo, Stephen A. Ellwood, David W. Macdonald, Andrew Markham, Cecilia Mascolo, Bence Pasztor, Salvatore Scellato, Niki Trigoni, Ricklef Wohlers, and Kharsim Yousef. Evolution and sustainability of a wildlife monitoring sensor network: the wildsensing system. *ACM Transactions on Sensor Networks,* 2011.

[7] Frans Ekman, Ari Keränen, Jouni Karvo, and Jörg Ott. Working Day Movement Model. In *Proc. of in 'Proceeding of MobilityModels '08.* ACM, 2008.

[8] M. C. Gonzalez, C. A. Hidalgo, and A.-L. Barabasi. Understanding individual human mobility patterns. *Nature,* 453, June 2008.

[9] W.J. Hsu, T. Spyropoulos, K. Psounis, and A. Helmy. Modeling time-variant user mobility in wireless mobile networks. In *IEEE INFOCOM,* May 2007.

[10] T. Karagiannis, J.-Y. L. Boudec, and M. Vojnovic. Power law and exponential decay of inter contact times between mobile devices. In *Proceedings of ACM MobiCom,* 2007.

[11] S.J. Kim, I. Rhee, S. Chong, K. Lee, and S. Hong. Slaw: A new mobility model for human walks. In *Proceedings of IEEE INFOCOM,* May 2007.

[12] M. Musolesi and C. Mascolo. A community based mobility model for ad hoc network research. In *ACM/SIGMOBILE REALMAN,* Florence, Italy, May 2006.

[13] H.A. Nguyen, S. Giordano, and A. Puiatti. Probabilistic routing protocol for intermittently connected mobile ad hoc network (propicman). In *IEEE International Symposium on a World of Wireless, Mobile and Multimedia Networks (WoWMoM),* pages 1–6, 2007.

[14] I. Rhee, M. Shin, S. Hong, K. Lee, and S. Chong. On the levy walk nature of human mobility. In *Proceedings of IEEE INFOCOM,* 2008.

[15] James Scott, Richard Gass, Jon Crowcroft, Pan Hui, Christophe Diot, and Augustin Chaintreau. CRAWDAD data set cambridge/haggle (v. 2009-05-29). Downloaded from http://crawdad.cs.dartmouth.edu/cambridge/haggle, may 2009.

[16] Thrasyvoulos Spyropoulos and Andreea Picu. *Opportunistic Routing,* volume Chaper in: Mobile Ad hoc networking: the cutting edge directions. Elsevier, 2012.

[17] J. Stefa and A. Mei. Swim: a simple model to generate small mobile worlds. In *Proceedings of IEEE INFOCOM,* 2009.

[18] Qunwei Zheng, Xiaoyan Hong, Jun Liu, and David Cordes. Agenda driven mobility modelling. *Int. J. Ad Hoc and Ubiquitous Computing,* 5:22–36, 2010.

[19] H. Zhu, Y. Zhu, M. Li, and M. Ni. Seer: Metropolitan-scale traffic perception based on lossy sensory data. In *Proceedings of IEEE INFOCOM,* 2009.

# Less is More: Long Paths do not Help the Convergence of Social-Oblivious Forwarding in Opportunistic Networks

Chiara Boldrini
IIT-CNR
Via G. Moruzzi 1
56124, Pisa, Italy
chiara.boldrini@iit.cnr.it

Marco Conti
IIT-CNR
Via G. Moruzzi 1
56124, Pisa, Italy
marco.conti@iit.cnr.it

Andrea Passarella
IIT-CNR
Via G. Moruzzi 1
56124, Pisa, Italy
andrea.passarella@iit.cnr.it

## ABSTRACT

Message delivery in opportunistic networks is substantially affected by the way nodes move. Given that messages are handed over from node to node upon encounter, the intermeeting time, i.e., the time between two consecutive contacts between the same pair of nodes, plays a fundamental role in the overall delay of messages. A desirable property of message delay is that its expectation is finite, so that the performance of the system can be predicted. Unfortunately, when intermeeting times feature a Pareto distribution, this property does not always hold. In this paper, assuming heterogeneous mobility and Pareto intermeeting times, we provide a detailed study of the conditions for the expectation of message delay to converge when social-oblivious forwarding schemes are used. More specifically, we consider different classes of social-oblivious schemes, based on the number of hops allowed, the number of copies generated, and whether the source and relay nodes keep track of the evolution of the forwarding process or not. Our main finding is that, as long as the convergence of the expected delay is concerned, allowing more than two hops does not provide any advantage. At the same time, we show that using a multi-copy scheme can, in some cases, improve the convergence of the expected delay.

## Categories and Subject Descriptors

C.2.1 [**Computer-Communication Networks**]: Network Architecture and Design—*Wireless Communication*; C.2.2 [**Computer-Communication Networks**]: Network Protocols—*Routing protocols*

## Keywords

opportunistic networks, forwarding protocol, expected delay convergence

## 1. INTRODUCTION

The great popularity of the delay tolerant networking paradigm is due to its ability to cope with challenged network conditions, such as high node mobility, variable connectivity, and disconnected subnetworks, that would impair communications in traditional Mobile Ad Hoc Networks. Opportunistic networks are an instance of the delay tolerant paradigm applied to networks made up of users' portable devices (such as smartphones and tablets). In this scenario, user mobility becomes one of the main drivers to enable message delivery. In fact, according to the store-carry-and-forward paradigm, user devices store messages and carry them around while they move in the network, exchanging them upon encounter with other nodes, and eventually delivering them to their destination.

An opportunistic forwarding protocol defines the strategy according to which messages are exchanged during encounters. Two main approaches can be identified. On the one hand, there are *social-oblivious* protocols, which do not exploit any information about the users' context and social behaviour but just hand over the message to the first node encountered (avoiding at most those nodes that have already forwarded the message). The main advantage of these strategies is that they are intrinsically simple and lightweight (practically no information to collect, store, or mine). This simplicity, however, is typically paid in terms of suboptimal routing performance. In order to improve message forwarding, smarter strategies have been proposed that exploit information on the social context users operate in. These approaches, referred to as *social-aware*, typically make use of information on how users behave or which social relations they share in order to make predictions on users' future behavior that might be useful for forwarding messages. Depending on the number of copies generated for the same message, forwarding protocols can be classified into single-copy or multi-copy schemes. In the first case, at any time, in the network there is just one copy of the message to be delivered, while in the second case more copies are generated, hoping that at least one of them will eventually reach the destination. Multi-copy strategies have been shown to improve the reliability of delivery with respect to single-copy approaches [13]. Forwarding protocols may also differ in the number of relays that they exploit. Simpler strategies may be single hop or two hops strategies (e.g. Direct Transmission and Two Hop [7]), while others can allow multi-hop paths to bring the message to the destination.

Modelling the performance of social-oblivious and social-aware forwarding protocols for opportunistic networks is still an open research issue. As messages follow multi-hop paths across the nodes of the network, their delay is the result of the delay accumulated at each hop along the forwarding path. Therefore, the time (*intermeeting time*) between consecutive encounters of a pair of nodes is the elementary component of the overall delay. Thus, knowing the distribution of intermeeting times, one could - in principle - model the distribution of the delay experienced by messages. Unfortunately, there is no agreement on the actual shape featured by pairwise intermeeting times in real networks. Of the many hypotheses that have been made [5] [6] [12][10], the most challenging from the forwarding standpoint is the one proposed by Chaintreau et al. [4]. Chaintreau et al. found intermeeting times extracted from real mobility traces to follow a Pareto[1] distribution. The problem with Pareto distributions is that their expectation is finite only for certain values of their exponent $\alpha$. More specifically, the expectation is finite if $\alpha > 1$, while for $\alpha \leq 1$ it diverges to infinity. Being the delay the result of the composition of the time intervals between node encounters, depending on the exponent values featured by intermeeting times, the expectation of the delay might diverge. Clearly, having a finite expected delay is a key requirement for any communication protocol.

Due to the relevance of the problem, in this paper, considering heterogeneous mobility patterns, we derive the conditions on the Pareto exponent of pairwise intermeeting times under which the expectation of the delay under multi-copy and/or multi-hop social-oblivious protocols is finite. The starting point of our paper is the work by Chaintreau et al. [4], where such conditions have been studied for the the single-copy two-hop scheme and flooding (see Section 2 for more details) under the assumption of homogeneous mobility. Homogeneous mobility implies that the intermeeting times between any pair of nodes have the same statistical characteristics (e.g., same exponent for Pareto intermeeting times). Recent works have shown, however, that real networks are intrinsically heterogeneous [5]. In this paper, we investigate whether heterogeneity in contact patterns helps the convergence of the expected delay of a general class of social-oblivious forwarding protocols and whether convergence conditions can be improved using multi-copy strategies and/or multi-hop paths.

We anticipate here that longer paths, i.e., with a number of allowed hops greater than two, do not help the convergence of social-oblivious strategies. The intuitive explanation is that two hops are enough for exploring the forwarding diversity available in the network. In fact, the relay to which the source hands over the message under the two-hop scheme can be any other node in the network, just as in the multi-hop case in which the number of allowed hop is greater than two. On the other hand, we find that multi-copy strategies can achieve a finite expected delay even when single-copy strategies cannot. This is due to the fact that a parallel delivery of more than one copy can increase the chances of finding the destination.

The paper is organised as follows. In Section 2 we briefly review the state of the art on forwarding protocols for opportunistic networks. In Section 3 we describe the network model we consider and the assumptions we make. Then, in Section 4 we identify the main techniques that can be applied to social-oblivious forwarding, thus identifying a set of representative classes of social-oblivious schemes. For these classes, we derive in Section 5 the conditions for the expectation of their delay to be finite. Finally, Section 6 concludes the paper.

---

[1]In the following we use the terms "Pareto" and "power law" interchangeably.

## 2. RELATED WORK

As discussed above, forwarding protocols for opportunistic networks can be classified as social-oblivious or social-aware protocols, depending on whether they use information on the way nodes behave in order to make forwarding decisions. In the following we only consider social-oblivious schemes, as they are the focus of this work. The simplest social-oblivious protocol is Direct Transmission [7], in which the source node is only allowed to deliver the message directly to the destination, if ever encountered. At the opposite side of the spectrum, with Epidemic routing [14] a new copy of the message is generated and handed over (both by the source and intermediate relays) any time a new node is encountered. In an ideal scenario without resource limitations Epidemic achieves the minimum possible delay, but in realistic settings it is typically unfeasible due to the huge amount of resources it consumes [13]. In order to mitigate the side effects of Epidemic-style forwarding schemes in resource constrained environments, controlled flooding solutions have been proposed (e.g., Spray&Wait [13], gossiping [8]). Another popular social-oblivious forwarding protocol is the Two Hop scheme [7], in which a message is forwarded by the source node to the first node encountered, which is then allowed only to pass the message directly to the destination. The Two Hop strategy has been shown to guarantee the maximum throughput capacity in a homogeneous network [7].

To the best of our knowledge, there is no other contribution besides that of Chaintreau et al. [4] that considers the problem of the convergence of the expected delay when intermeeting times feature a Pareto distribution. Our work differs from that of Chaintreau et al. both in the mobility settings and in the forwarding schemes considered. More specifically, we focus on the more realistic case of heterogeneous intermeeting times (as opposed to the homogeneous mobility considered in [4]) and we extend the set of social-oblivious policies considered. As a check of correctness, in Section 5 we apply our derivation to the homogeneous case and, under the same configuration used by Chaintreau et al. in [4], we obtain the same results.

## 3. NETWORK MODEL

Our model considers a network with $N$ mobile nodes. We denote with $\mathcal{N}$ the set of all nodes in the network. For the sake of simplicity, we hereafter assume that messages can be exchanged only at the beginning of a contact between a pair of nodes and that the transmission of the relayed messages can be always completed within the duration of a contact. In addition, we assume that each message is a bundle [3], an atomic unit that cannot be fragmented. We also assume infinite buffer space on nodes. All the above assumptions allow us to isolate, and thus focus on, the effects of node mobility from other effects, and are common assumptions in the literature on opportunistic networks modelling (they are used in most of the literature reviewed in Section 2).

Given that messages are handed over from node to node before reaching their destination, the way nodes move heavily affects the delay experienced by messages. As we assume that the transmission of a message can always be completed during a pair-wise contact, the actual duration of the contact is not critical. Thus, the main role in the experienced delay is played by intermeeting times, which are defined as follows.

DEFINITION 1 (INTERMEETING TIME).
*The intermeeting time $M_{ij}$ between node $i$ and node $j$ is defined as the time between two consecutive meetings between the same pair of nodes. If $t_f$ is the time at which a contact between node $i$ and node $j$ has just finished, the intermeeting time $M_{ij}$ is given by:*

$$M_{ij} = \min_{t > t_f} \{ t - t_f : ||X_i(t) - X_j(t)|| < r \} \qquad (1)$$

*where $X_i(t)$ and $X_j(t)$ denote the position of $i$ and $j$ at time $t$, and $r$ is the transmission range[2].*

For the sake of tractability, we assume that intermeeting times between every specific node pair $i, j$ are independent and identically distributed and that their expectation $E[M_{ij}]$ does not vary with time (in other words, we assume a stationary network). By definition, the rate of encounter between node $i$ and node $j$ is given by $\frac{1}{E[M_{ij}]}$.

The message generation process and the mobility process are independent. Thus, the time at which a new message is generated can be treated as a random time in the evolution of the mobility process, and thus the message sees the network as an observer arriving at a random point in time would. For this reason, in our analysis we will often use the concept of residual intermeeting time.

DEFINITION 2 (RESIDUAL INTERMEETING TIME).
*Assuming that node $i$ and node $j$ are not in contact at time $t_o$, the residual intermeeting time $R_{ij}(t)$ between them is given by the time interval between $t_o$ and the first time node $i$ and node $j$ come into each other's range again, i.e.:*

$$R_{ij} = \min_{t > t_o} \{ t - t_o : ||X_i(t) - X_j(t)|| < r \}, \qquad (2)$$

*where $X_i(t)$ and $X_j(t)$ denote the position of $i$ and $j$ at time $t$, and $r$ is the transmission range.*

Under our assumption of Pareto intermeeting times, the intermeeting time $M_{ij}$ between a generic pair of nodes $i$ and $j$ is described by the following CCDF:

$$F_{M_{ij}}(t) = \left( \frac{t_{min_{ij}}}{t + t_{min_{ij}}} \right)^{\alpha_{ij}} \qquad (3)$$

in which we use the definition of the Pareto distribution which allows for values arbitrarily close to zero, usually denoted as American Pareto [11] [1] (as opposed to the European Pareto version). Parameters $\alpha_{ij}$ and $T_{min_{ij}}$ are usually referred to as the shape and scale of the Pareto distribution, respectively. Note that we do not require intermeeting times $M_{ij}$ and $M_{ji}$ to be symmetric. Please note also that being the American Pareto a European Pareto shifted by $t_{min_{ij}}$ to the left, both Pareto definitions share the same requirements for their expectation to converge. Thus, the following remark holds.

REMARK 1. *The Pareto distributions introduced above are defined for $\alpha_{ij} > 0$ (due to the required PDF normalization), and their expectation converges (i.e., is finite) when $\alpha_{ij} > 1$.*

---

[2]Without loss of generality, here we assume a deterministic unit disk graph model for radio propagation. In other words, nodes can communicate only if their current distance is smaller than the transmission range. This is a common assumption in the literature on opportunistic networks. The proposed framework still applies for every other model of radio propagation.

| $N$ | number of nodes in the network |
|---|---|
| $F_X$ | complementary cumulative distribution function (CCDF) of random variable $X$ |
| $X(x)$ | probability density function of random variable $X$ |
| $M_{ij}$ | intermeeting time for the $i, j$ node pair |
| $R_{ij}$ | residual intermeeting time for the $i, j$ node pair |
| $\alpha_{ij}$ | exponent (*shape*) of the Pareto distribution that characterises $M_{ij}$; we assume $\alpha_{ij} > 1, \forall i, j$ |
| $t_{min}$ | scale of the Pareto distribution that characterises $M_{ij}$; $t_{min} > 0$ |
| $D_i^d$ | delay of a message generated by node $i$ and addressed to node $d$ |
| $\mathcal{N}$ | set comprising all nodes of the network |
| $\mathcal{P}_i$ | set comprising all nodes that can be encountered by node $i$ |
| $h_{max}$ | maximum number of hops allowed |

**Table 1: Notation**

As we have already discussed, residual intermeeting times come into the picture more often than intermeeting times, because the time of the generation of new messages can be modelled as a random time with respect to the evolution of the mobility process. Following a standard approach [1], from an American Pareto random variable with shape $\alpha_{ij}$ and scale $t_{min_{ij}}$ we obtain residuals that feature an American Pareto distribution with shape $\alpha_{ij} - 1$ and scale $t_{min_{ij}}$. In the case of European Pareto, the residual is not exactly Pareto distributed but it converges to a Pareto distribution with shape $\alpha_{ij} - 1$ in the tail [1]. Thus, it shares the same convergence conditions as the residual of an American Pareto random variable. For the residual intermeeting time, the following remark holds.

REMARK 2. *The Pareto distribution of $R_{ij}$ is defined for $\alpha_{ij} > 1$ (due to the required PDF normalization), and its expectation converges when $\alpha_{ij} > 2$.*

The notation used throughout the paper is summarised in Table 1. Similarly to the reference literature [4][9], for ease of computation in the following we restrict to the case of power law random variables having the same scale, i.e., $t_{min_{ij}} = t_{min}, \forall i, j \in \mathcal{N}$. In addition, for the sake of comparison with [4], we also assume that the probability that two nodes meet is greater than zero for all node pairs. This ensures that, in principle, all nodes can meet with each others. Therefore, cases of deadlock (a message reaches a node which is impossible to leave due to the total absence of contacts with either other possible relays or the destination) are not possible. The only cause of divergent expected delay are therefore the distributions of intermeeting times.

## 4. FORWARDING STRATEGIES

In this section we summarise the main variants of opportunistic forwarding schemes that will be later evaluated against each other as far as the convergence of their expected delay is concerned. We identify three main strategies that forwarding protocols can adopt in order to improve their forwarding performance, namely the number of hops allowed, the number of copies generated, and whether the source and relay nodes keep track of the evolution of the forwarding process or not.

First, forwarding strategies can be single-copy or multi-copy. In the former case, at any point in time there can be at most one copy of each message circulating in the network. In the latter, multiple copies can travel in parallel, thus in principle multiplying the opportunities to reach the destination. These multiple copies can be all created and handed over by the source node, or also intermediate relays could be allowed to take part into the multiplication process. Here we only focus on source generation. Other possible configurations (e.g., intermediate relays allowed to generate new copies, like in the Spray&Wait case [13]) are left as future work.

Second, forwarding protocols can be classified based on the number of hops that they allow messages to traverse. In principle, this number could also be infinite. However, being such an approach not feasible in practice, the number of hops is either limited arbitrarily (e.g., using the TTL field) or is naturally constrained by the forwarding strategy (e.g., if each possible relay can be exploited just once, messages cannot perform more than $N-1$ hops). When the number of allowed hops is finite, the last relay can only deliver the message to the destination directly.

Third, the amount of knowledge that each agent in the forwarding process can rely on (or is willing to collect and store) is an additional element for classifying forwarding strategies. Focusing on the source node, there can be social-oblivious strategies in which the source node does not keep track at all of how the forwarding process progresses. In this case, considering the configuration in which the source node can generate up to $m$ copies of the message, the $m$ copies might end up being all distributed to the exact same relay, thus eliminating the potential benefits of multi-copy forwarding. A memoryful source, instead, is able to guarantee to use distinct relays. A similar problem holds for intermediate relays. Memoryless relays can forward the message to the same next hop more than once, because they are not at all aware of what happened in the past. On the other hand, memoryful relays possess this knowledge, and are able to refuse the custody of messages that they have already relayed. Please note that we assume that the source node can never be handed over messages that it has generated. This assumption simply takes into account the fact that the source identity is always enclosed into the message header, thus this does not require any additional knowledge beside what is already present in the system.

Table 2 summarizes the feasible combinations (the ones marked with the checkmarks) of the forwarding characteristics described above when social-oblivious schemes are considered. These combinations can be found in well known routing strategies. For example, the 1-hop 1-copy memoryless forwarding corresponds to the Direct Transmission strategy [7], in which the source node can only deliver the messages to the destination. The 2-hop 1-copy memoryless forwarding is equivalent to the Two Hop forwarding introduced in [7]. The 2-hop $m$-copy memoryful forwarding is equivalent to the multi-copy version of the Two Hop protocol studied in [4]. Please note that relays can be memoryful only when they have multiple forwarding choices. This is not the case when the number of hops is limited to either one (there is no relay in this case) or two (relays can only deliver the message to the destination).

# 5. EXPECTED DELAY CONVERGENCE FOR SOCIAL-OBLIVIOUS SCHEMES

In this section we study under which conditions the expected delay of the social-oblivious schemes described in Section 4 converges for a tagged source-destination pair. Simultaneous convergence for all source-destination pairs would require combining the conditions derived in the paper, but the problem is not touched upon due to lack of space. Recall that according to social-oblivious forwarding a message is handed over to the first feasible relay encountered. In the following, we denote with $\mathcal{P}_i$ the set of all nodes that can be encountered by node $i$ (i.e., the probability of an encounter with node $i$ is strictly greater than zero). Recall, also, that we assume that $\alpha_{ij} > 1$ for all $i, j$ node pairs, so that the residual inter-meeting times are defined (see Remark 2). It is easy to show that, when $\alpha_{ij} \leq 1$, none of the forwarding algorithms considered in this paper are able to achieve a convergent expected delay. We refer the interested reader to [2] for the complete proof.

## 5.1 Single-copy schemes

In a previous work we have studied the single-copy case for both the 1-hop and 2-hop social oblivious forwarding protocols. For the readers' convenience, we hereafter recall these findings, whose proofs can be found in [2].

THEOREM 1 (SINGLE-COPY ONE-HOP SCHEME). *In a heterogeneous network where the intermeeting time $M_{ij}$ between any generic $i, j$ node pair follows a power law distribution with shape $\alpha_{ij}$, when the Single-copy One Hop relaying protocol (also known as Direct Transmission protocol) is used the expected delay for messages generated by the source node $s$ for the destination node $d$ converges if and only if $\alpha_{sd} > 2$.*

THEOREM 2 (SINGLE-COPY TWO-HOP SCHEME). *In a heterogeneous network where the intermeeting time $M_{ij}$ between any generic $i, j$ node pair follows a power law distribution with shape $\alpha_{ij}$, when the single-copy two-hop relaying protocol is used, the expected delay for messages generated by the source node $s$ for the destination node $d$ converges if and only if both the following conditions hold true:*

**C1** $\sum_{j \in \mathcal{P}_s} \alpha_{sj} > 1 + |\mathcal{P}_s|$, *where $\mathcal{P}_s$ denotes the set of all nodes that can be encountered by node $s$;*

**C2** $\alpha_{jd} > 2$, $\forall j \in \mathcal{P}_s - \{d\}$.

According to Theorem 1, the Direct Transmission protocol yields a convergent expected delay only if the source node meets the destination with a residual intermeeting time whose expectation converges. This clearly follows from the fact that the source node cannot exploit any other relays for the forwarding of the message. In the case of the two-hop scheme, the expectation converges even if the source node is not able to ensure convergence with a direct delivery. This can happen if the source node is able to hand over the message to any of the possible relays within a convergent expected time (Condition C1) and if the meeting process between this relay and the destination has a residual whose expectation converges (Condition C2). Please note that condition C1 alleviates the convergence condition on the source node at the expense of the additional condition C2 on intermediate relays.

With Theorem 3 we extend the analysis of single-copy schemes by studying their $n$-hop version. Recall that, as shown in Table 2, with the $n$-hop single-copy social-oblivious forwarding we must consider both the memoryless and the memoryful case for relays. Thus, in the memoryless case, relays hand over the message to the first encountered node,

| | 1 hop | | 2 hops | | $n$-hop | |
|---|---|---|---|---|---|---|
| | 1 copy | $m$ copies | 1 copy | $m$ copies | 1 copy | $m$ copies |
| memoryless | ✓ | - | ✓ | ✓ | ✓ | ✓ |
| memoryful source | - | - | - | ✓ | - | ✓ |
| memoryful relays | - | - | - | - | ✓ | ✓ |

Table 2: Summary of social-oblivious routing strategies

regardless of whether this node has already relayed the message or not. On the other hand, memoryful relays guarantee that the message is relayed at most once by each node.

THEOREM 3 (SINGLE-COPY $n$-HOP SCHEME).
*In a heterogeneous network where the intermeeting time $M_{ij}$ between any generic $i, j$ node pair follows a power law distribution with shape $\alpha_{ij}$, when the single-copy $n$-hop relaying protocol (both in the memoryless and memoryful case) is used, the expected delay for messages generated by the source node $s$ for the destination node $d$ converges if and only if conditions C1 and C2 in Theorem 2 hold true.*

PROOF. Here we only provide a sketch for the proof, whose complete version can be found in the associated technical report [2]. The proof is composed of three parts. We first study the delivery from the source node to the relay, then we concentrate on the delivery from relay to relay along the multi-hop path, and finally we study the delivery from the last relay to the destination node.

The source node $s$ can either deliver the message directly to the destination or hand it over to an intermediate relay. Recall that we model message arrival time as a random point in time with respect to the evolution of the mobility process. Thus, the time before the source node releases the message is distributed as $\min_{j \in \mathcal{P}_s}\{R_{sj}\}$, which is the time before the first node (possibly including the destination) is encountered. It can be showed that $\min_{j \in \mathcal{P}_s}\{R_{sj}\}$ features a Pareto distribution with shape $\sum_{j \in \mathcal{P}_s}(\alpha_{sj} - 1)$, which, according to Remark 2, should be greater than 1 for the expectation to converge. This implies $\sum_{j \in \mathcal{P}_s} \alpha_{sj} > 1 + |\mathcal{P}_s|$, thus obtaining condition C1.

Once the source node has handed over the message, we know that the message will follow a $n$-hop path, with $n \leq \min\{N-1, h_{max}\}$ for the memoryful case and $n \leq h_{max}$ in the memoryless case. First, note that any node $i \in \mathcal{N} - \{s, d\}$ has a non negligible probability of being the $k$-th hop along the $n$-th hop path, with $k \in \{1, ..., n-1\}$. In fact, assume for a moment that the message can leave any node within a finite expected delay (conditions under which this assumption is true are provided below). Then, given that we assume $\alpha_{ij} > 1$ for all $i, j$ node pairs, i.e., that nodes can meet with any other node, at each forwarding step every node has a non negligible probability of being selected.

Let us now derive the conditions for the expected time before the message leaves a node to be finite. Before considering intermediate relays, let us focus on the delivery from the last relay to the destination node. It is possible to prove that the delivery from the last relay to the destination shares the same convergence condition on its expectation as the residual intermeeting time between the relay and the destination. From Remark 2 we know that $R_{jd}$ has finite expectation if $\alpha_{jd} > 2$. Given that all nodes have a non negligible probability of being the $(n-1)$-th hop, as we proved above, condition $\alpha_{jd} > 2$ must be satisfied for all nodes $j \in \mathcal{N} - \{s, d\}$. Under our assumption of nodes all potentially meeting with each other, this condition is equivalent to condition C2, as $\mathcal{N} - \{s, d\} = \mathcal{P}_s$.

In order to complete the proof we should also derive the conditions under which the delivery from intermediate relay to intermediate relay achieves finite expected delay. This derivation is quite involved, thus we left it entirely to the associated technical report. Note, however, that, given that conditions C1 and C2 are required for the first and last hop, the overall convergence conditions for the single-copy $n$-hop scheme can at most be equal to those of the single-copy 2-hop scheme, not better. Specifically, in [2] we derive that condition C2, that must apply to all nodes in $\mathcal{N} - \{s, d\}$, guarantees that the conditions derived for intermediate relays are automatically satisfied, both for the memoryless and the memoryful case. Thus, overall, conditions C1 and C2 guarantee the convergence of the expected delay under then $n$-hop single-copy social-oblivious scheme. □

Theorem 3 tells us that, when using single-copy social-oblivious schemes, letting the message traverse more than two hops does not improve the convergence of the expected delay. Thus, when convergence is the only goal, network resources can be saved using the two-hop social-oblivious scheme without impairing convergence of the expected delay.

## 5.2 Multi-copy schemes

As discussed in Section 2, when multiple copies of the same message can travel in parallel, the opportunities to reach the destination are multiplied. In this section we investigate whether this also positively affects the convergence of the expected delay. In the following we present the results for the convergence of multi-copy/multi-hop schemes, while a detailed discussion of their advantages and disadvantages with respect to the single-copy schemes analyzed above will be provided in Section 5.3. Please also note that hereafter we only provide an intuitive sketch for the proofs, which can be found in a detailed version in the associated technical report [2].

### 5.2.1 Two-hop forwarding

Recall that, according to the multi-copy version of the two-hop forwarding scheme, the source node hands over a copy of the message to the first $m$ encountered nodes, which will then be only allowed to deliver the message directly to the destination, if ever met. Moreover, in the *memoryless* case, the source node does not keep a record of the relay nodes used so far, and thus two consecutive encounters with the same node will end up in the message being copied again to the same relay. In the *memoryful* case, a relay node cannot be used more than once. As we discuss below, these different capabilities have a great impact on the convergence of the expected delay.

THEOREM 4 ($m$-COPY MEMORYLESS TWO-HOP). *In a heterogeneous network where the intermeeting time $M_{ij}$ between any generic $i, j$ node pair follows a power law distribution with shape $\alpha_{ij}$, when the memoryless multi-copy two-hop relaying protocol is used, the expected delay for messages generated by the source node $s$ for the destination node* *d converges if and only if conditions C1 and C2 in Theorem 2 hold true.*

PROOF. The proof is composed of two parts. First, we discuss the convergence conditions for the $m$ copies sent by the source node. The first copy can be studied analogously to what we did in Theorem 3, thus obtaining condition C1. For the following copies, we have to consider that relays can be reused, thus the time before the $k$-th copy leaves the source is given by the minimum of both residual intermeeting times (for nodes that have not been yet used as relays) and intermeeting times (for nodes already used as relays), because the mobility process regenerates upon encounter. We prove that the fact that the delivery of the $k$-th copy is constrained to start after all previous copies have been delivered does not affect convergence for the mobility scenario considered in the paper and, for the sake of clarity, here we neglect it (it is, however, considered in [2]). Thus, we have that, every time a new copy is handed over, a residual intermeeting time in the initial set $\{R_{sj}\}_{j \in \mathcal{P}_s}$ from which we take the minimum is substituted with the corresponding intermeeting time. Given that convergence conditions are looser for intermeeting times than for residual intermeeting times (see Remarks 1 and 2), it follows that condition C1, which is the necessary and sufficient convergence condition that applies to the case in which there are only residuals, is also a sufficient condition for the cases in which intermeeting times and residual intermeeting times are mixed.

The analysis of the second hop (delivery from relays to the destination) starts from the consideration that, given that the protocol is memoryless, the number of distinct relays actually carrying a copy of the message ranges from 1 to $m$. The worst case from the convergence standpoint is that in which all copies have been relayed to the same node. As this can happen with a non negligible probability, this worst case would impair convergence and thus we have to avoid it ensuring that condition C2 holds for all possible relays. □

In the following we derive the convergence conditions for the expected delay under the memoryful $m$-copy two-hop scheme. To this aim, in Lemma 1 we prove the existence of an operating point $m^*$ for the memoryful $m$-copy two-hop scheme such that, when $m \leq m^*$, the expected time before all $m$ copies are delivered to their $m$ relays converges, while for $m > m^*$ copies exceeding $m^*$ never achieve a convergent expected delay. Please recall that in this paper we assume $\alpha_{ij} > 1$ for all $i, j$ node pairs.

LEMMA 1. *In a heterogeneous network where the intermeeting times $M_{ij}$ between any generic $i, j$ node pair follow a power law distribution with shape $\alpha_{ij}$ and the memoryful $m$-copy two-hop forwarding protocol is in use, there exists a characteristic value $m^*$ such that, when $m \leq m^*$, the expected time before all $m$ copies are delivered to their $m$ relays converges, while for $m > m^*$ copies exceeding $m^*$ never achieve a convergent expected delay. The value of $m^*$ can be obtained as follows:*

$$m^* = \begin{cases} 0 & \text{if } \sum_{j \in \mathcal{P}_s} \alpha_{sj} \leq N \\ \arg\max_m \{ m + \sum_{i=m}^{N-1} \alpha_i^* > 1 + N \} & \text{o.w.} \end{cases} , \quad (4)$$

*where $\alpha_i^*$ denotes the $i$-th largest $\alpha_{sj}$ with $j \in \mathcal{P}_s$.*

PROOF. We consider a memoryful source that is delivering the $m$ initial copies of the message. As the source is memoryful, after the $k$-th copy is relayed, the next copy can

be delivered to the subset of nodes that comprises only those that have not been already used as relay. We prove that the convergence conditions become stricter as the cardinality of the set from which we choose the relays decreases. This let us focus on the delivery of the $m$-th copy, because that is the one that sees the smallest set of possible relays, whose cardinality is $N - m$. Following the same line of reasoning discussed in the proof of Theorem 3, we prove that all nodes have a non negligible probability of being chosen as relays at each step. Thus the sets of possible relays are given by all possible combinations of $N - m$ relays taken from the initial $N - 1$. The worst combination, as far as the convergence of the expected delay of the $m$-th copy is concerned, is that containing the $N - m$ nodes having the smallest Pareto exponent $\alpha_{sj}$, with $j \in \mathcal{P}_s$. If we show that the $m$-th copy is handed over by the source node in a finite expected time, then the convergence for all previous copies and all cases different from the worst one will automatically follow. We derive that the $m$-th copy is relayed within a finite expected time if the sum of the $N - m$ smallest exponents $\alpha_{sj}$ with $j \in \mathcal{P}_s$ is greater than $1 + N - m$. In order to find the value $m^*$ corresponding to the maximum number of copies that can be sent within a finite expected time, we simply compute the maximum $m$ value such that the above condition is satisfied. □

Finally, in Theorem 5 we provide the convergence conditions for the overall expected delay under the memoryful $m$-copy two-hop scheme operating at $m \leq m^*$.

THEOREM 5 ($m$-COPY MEMORYFUL TWO-HOP). *In a heterogeneous network where the intermeeting times $M_{ij}$ between any generic $i, j$ node pair follow a power law distribution with shape $\alpha_{ij}$, when the memoryful $m$-copy two-hop forwarding protocol operating at $m \leq m^*$ is used, the expected delay for messages generated by the source node $s$ for the destination node $d$ converges if and only if $\sum_{j=N-m}^{N-1} \alpha_j' > 1 + m$, where $\alpha_j'$ denotes the $j$-th largest $\alpha_{jd}$ with $j \in \mathcal{P}_s$ (thus $\sum_{j=N-m}^{N-1} \alpha_j'$ is the sum of the $m$ smallest $\alpha_{jd}$ with $j \in \mathcal{P}_s$).*

PROOF. The proof focuses on the second hop, as the delivery from source to relay is guaranteed to have finite expected delay by condition $m < m^*$. The second hop can be modelled as a parallel delivery from the $m$ relays to the destination. For each of the relays, the residual intermeeting time with the destination starts from when the message has been received by the relay. In the worst case, the set of relays currently holding a copy of the message is composed by nodes having the lowest exponent value for the intermeeting time with the destination. Starting from these considerations and using Remark 2 and some Pareto properties derived in [2], we prove condition $\sum_{j=N-m}^{N-1} \alpha_j' > 1 + m$. □

As discussed before, Chaintreau et al. [4] studied the $m$-copy memoryful two-hop scheme under homogeneous mobility patterns (corresponding to $\alpha_{ij} = \alpha, \forall i, j$). For the sake of completeness, in Corollary 1 we verify that Theorem 5 confirms and extends the results in [4].

COROLLARY 1. *In a homogeneous network where the intermeeting times $M_{ij}$ follow a power law distribution with shape $\alpha$ for all $i, j$ node pairs, when the $m$-copy two-hop strategy ($m \leq m^*$) is used, the expected delay for messages generated by the source node $s$ for the destination node $d$ converges if and only if*

$$\alpha > \frac{1}{N-m} + 1. \qquad (5)$$

*In addition, $m^*$ is given by $m^* = \left\lfloor N - \frac{1}{\alpha-1} \right\rfloor$.*

Please note that the necessary and sufficient condition in Equation 5 extends the sufficient condition provided by Chaintreau et al. [4]. In fact, Chaintreau et al., under the assumption $N > 2m$ (which we have relaxed), derive that the expected delay of the $m$-copy two-hop scheme ($m \le m^*$) converges in a homogeneous setting as long as $\alpha > 1 + \frac{1}{m}$. Exploiting assumption $N > 2m$, we have that $N - m > m$, thus $\frac{1}{N-m} < \frac{1}{m}$, and $1 + \frac{1}{N-m} < 1 + \frac{1}{m}$. Thus, when condition $\alpha > 1 + \frac{1}{m}$ is verified, also Equation 5 holds true.

### 5.2.2  Multi-hop forwarding

Again we consider a social-oblivious protocol in which the source node generates $m$ copies of the message and hands them over to the first $m$ nodes encountered. Once the source node has handed over the $m$ copies, the message travels along multi-hop social-oblivious paths until the destination is found. Based on the type of memory applied to the source node and to the relays, we consider the following versions of the $n$-hop $m$-copy protocol (corresponding to the last column of Table 2):

**V1** the source node does not keep track of already used relays nor the intermediate relays do

**V2** the source node selects $m$ distinct nodes but the intermediate relays are not aware of already used relays

**V3** the source node selects $m$ distinct nodes and the intermediate relays can relay the message only once.

Theorem 6 describes the convergence conditions that apply in all these cases.

THEOREM 6  ($m$-COPY $n$-HOP). *In a heterogeneous network where the intermeeting time $M_{ij}$ between any generic $i, j$ node pair follows a power law distribution with shape $\alpha_{ij}$, when either the V1, V2, or V3 social-oblivious $m$-copy $n$-hop protocol is used, the expected delay for messages generated by the source node $s$ for the destination node $d$ converges if and only if condition C1 and C2 in Theorem 2 hold true.*

PROOF. Due to lack of space, here we only discuss the V3 case, as, being it memoryful, it is expected to perform better than memoryless social-oblivious forwarding. Proofs for V1 and V2 forwarding can be found in [2]. As we did before, here we only sketch the proof and we refer the reader to the associated technical report for the rigorous mathematical derivation.

With V3 forwarding, any time a copy of the message is handed over (either by the source or by an intermediate relay) a relay is removed from the set of possible relays. We identify a worst case (which we show to happen with non negligible probability) in which the first copy overtakes all other copies. This is the case in which the source node is not able to hand over the second copy because the first one has already used all possible relays. Thus, in the worst case V3 forwarding becomes a single-copy multi-hop forwarding, for which Theorem 3 holds. Please note that in V3 forwarding Lemma 1 does not hold due to the overtaking effect.  □

### 5.3  Discussion

Table 3 summarises the results derived so far for social-oblivious forwarding protocols. The first interesting finding is that $n$-hop social-oblivious protocols (last two columns of Table 3) are no more effective in delivering the message with finite expected delay than the simple 1-copy 2-hop forwarding. In fact, both $n$-hop social-oblivious protocols and the 1-copy 2-hop scheme share the same convergence conditions (C1 and C2), but the former consumes much more network resources than the latter. This tells us that, if we are only interested in the convergence of the expected delay, paths with more than two hops should be avoided, as two hops ensure that the available forwarding diversity between nodes is explored, while minimizing resource consumption.

With social-oblivious protocols, when the source node meets the destination with a residual intermeeting time having $\alpha_{sd} > 2$, there is no reason to exploit other relays, as this will only introduce the chance of picking a bad relay. This is confirmed by the fact that when the number of hops is allowed to grow, we have to impose on intermediate relays additional constraints that are not needed by Direct Transmission (see, e.g., condition C2 in Theorem 3 which requires that the residual intermeeting time between any relay and the destination achieves a finite expectation).

Different is the situation in which $\alpha_{sd} \le 2$. In this case, the source node is not able to directly deliver the message within a finite expected time, and thus exploring more relays is convenient as it allows the source node to exploit node diversity. In fact, even if the source node cannot reach destination $d$ directly with a finite expected delay, it may be able to hand over the message to other nodes within a finite expected time. If these intermediate relays are all able to individually deliver the message to the destination within a finite expected time, then the 1-copy 2-hop strategy guarantees convergences while minimizing resource consumption.

When there exists at least one intermediate relay which is not able to deliver the message directly to the destination within a finite expected time, the most effective strategy is the $m$-copy 2-hop forwarding. In fact, with $m$-copy 2-hop forwarding the source is able to send up to $m^*$ copies of the message. In the worst case $m^* = 1$, and thus we find again conditions C1 and C2 that hold for the 1-copy 2-hop strategy. But if the source node can reach operating point $m^* > 1$, conditions on the delivery from the relays to the destination become less restrictive as condition C4 can tolerate exponents $\alpha_{jd}$ smaller than 2 (the exact tolerance depends on the actual value of $m^*$).

### 5.4  A case study

In this section we apply the convergence conditions derived above to a toy scenario that we generate in order to illustrate the main differences among the social-oblivious strategies as far as the convergence of their expected delay is concerned. More specifically, we focus on forwarding strategies with different convergence conditions, which, as shown in Table 3, are 1-hop 1-copy (Direct Transmission), 2-hop 1-copy, and 2hop $m$-copy (with memoryful source) schemes. We consider 10 nodes, and the following set of exponents:

$$\boldsymbol{\alpha} = \{2.1, 2, 1.9, 1.8, 1.7, 1.6, 1.5, 1.4, 1.3\},$$

whose components are denoted as $\alpha_1, ..., \alpha_{N-1}$. We assume that a generic node $i$ meets all other nodes in a way such that $\alpha_{i,1} = \alpha_1, ..., \alpha_{i,i-1} = \alpha_{i-1}, \alpha_{i,i+1} = \alpha_i, ..., \alpha_{i,N} = \alpha_{N-1}$. We also set $t_{min}$ to $1s$. Let us consider messages sent by source node 1 with destination node 10. According to the results of Section 5.3, in this case the expected delay for the Direct Transmission is not defined, because $\alpha_{1,10} = 1.3$, while it should be greater than 2 for convergence. Analogously, the convergence condition for the 1-copy 2-hop scheme is

| | 1 hop | | 2 hops | | $n$-hop | |
|---|---|---|---|---|---|---|
| | 1 copy | $m$ copies | 1 copy | $m$ copies | 1 copy | $m$ copies |
| memoryless | $\alpha_{sd} > 2$ | - | [C1,C2] | [C1,C2] | [C1,C2] | [C1,C2] |
| memoryful source | - | - | - | [C3,C4] | - | [C1,C2] |
| memoryful relays | - | - | - | - | [C1,C2] | [C1,C2] |

**Table 3: Summary of convergence conditions for social-oblivious routing strategies (C1 and C2 are defined in Theorem 2, $C3 = m \leq m^*$ and $C4 = \sum_{j=N-m}^{N-1} \alpha_j' > 1 + m$)**

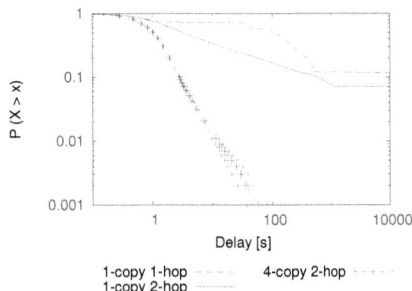

**Figure 1: Comparison of delay CCDFs for the 1-hop 1-copy, 2-hop 1-copy, and 2-hop $m$-copy schemes**

not satisfied. More specifically, condition C2 is not satisfied, because $\alpha_{i,d} < 2$ for all nodes $i$. The only scheme able to achieve a convergent expected delay is the $m$-copy 2-hop scheme. In fact, applying Lemma 1 we derive that the source can send up to $m^* = 4$ copies of the message for which the expectation of the latency is defined. If we assume to send all these four copies ($m = m^*$), condition C4 in Table 3 becomes $\sum_{j=6}^{9} \alpha_j > 5$. Given that the sum of the four smallest exponent in $\alpha$ is equal to 5.8, condition C4 is satisfied.

In order to complement these results, we ran a set of simulations, using a custom simulator written in C++, in which node 1 sends messages to node 10 according to a Poisson process with mean 1 second. In order for the comparison to be fair, we run $20000s$ of simulated time and we considered only the messages generated in the first $10000s$ in our statistics. The $10000s$ packet lifetime has been chosen in order to be significantly greater than the expected delay ($\sim 2s$) from node 1 to node 10 when the 4-copy 2-hop scheme is used. Please recall that the 4-copy 2-hop protocol is the only social-oblivious scheme to achieve a finite expected delay in this scenario. When applicable, i.e., when the average value is finite, we also show the 99% confidence intervals. For the three forwarding strategies discussed above, we plot the empirical cumulative distribution function in Figure 1. As expected, in the case of 4-copy 1-hop scheme, the great majority of messages ($\sim 99.9\%$) is delivered within a short time ($100s$) from their generation. For both the 1-hop 1-copy and the 2-hop 1-copy schemes, instead, after 10000 seconds there is still a big fraction (around 10%) of messages to be delivered. These long delays, predicted by our model, are those that cause the expected delay to diverge.

## 6. CONCLUSIONS

Assuming heterogenous, Pareto distributed, intermeeting times, in this paper we have derived the conditions on the Pareto exponents such that the expected delay of a large family of forwarding protocols is finite. We have considered different classes of social-oblivious strategies based on the number of copies and the number of maximum relays that are allowed. Our main result is that convergence is not improved by an increased number of allowed hops. Specifically, there is no advantage, as far as the convergence of the expected delay is concerned, in using more than two hops. In addition, when the source node is able to directly deliver the message to the destination with a finite delay, any additional relay can only add more restrictive convergence conditions.

As for the comparison of single-copy and multi-copy schemes, we found that multi-copy strategies can, in some cases, outperform single-copy strategies in terms of convergence of the expected delay. More specifically, a multi-copy two-hop strategy can prove effective when neither the source node nor intermediate relays are able to directly deliver the message to the destination within a finite expected time. The use of multiple copies, in fact, benefits from the parallel delivery of the message from different nodes, which may overcome the individual limitations in achieving a finite expected delay.

## 7. ACKNOWLEDGMENTS

This work was partially funded by the European Commission under the SCAMPI (FP7-FIRE 258414), RECOGNITION (FP7 FET-AWARENESS 257756), and EINS (FP7-FIRE 288021) projects.

## 8. REFERENCES

[1] C. Boldrini, M. Conti, and A. Passarella. From pareto inter-contact times to residuals. *IEEE Commun. Lett.*, 15(11):1256 – 1258, 2011.

[2] C. Boldrini, M. Conti, and A. Passarella. Less is more: long paths do not help the convergence of social-oblivious forwarding in opportunistic networks. Technical report, IIT-CNR, 2011, http://cnd.iit.cnr.it/chiara/pub/mobiopp12_tr.pdf.

[3] S. Burleigh, A. Hooke, L. Torgerson, K. Fall, V. Cerf, B. Durst, K. Scott, and H. Weiss. Delay-tolerant networking: an approach to interplanetary internet. *IEEE Commun. Mag.*, 41(6):128–136, 2003.

[4] A. Chaintreau, P. Hui, J. Crowcroft, C. Diot, R. Gass, and J. Scott. Impact of human mobility on opportunistic forwarding algorithms. *IEEE Trans. Mobile Comput.*, pages 606–620, 2007.

[5] V. Conan, J. Leguay, and T. Friedman. Characterizing pairwise inter-contact patterns in delay tolerant networks. In *Autonomics'07*, 2007.

[6] W. Gao, Q. Li, B. Zhao, and G. Cao. Multicasting in delay tolerant networks: a social network perspective. In *ACM MobiHoc'09*, pages 299–308. ACM, 2009.

[7] M. Grossglauser and D. Tse. Mobility increases the capacity of ad hoc wireless networks. *IEEE/ACM Trans. on Netw.*, 10(4):477–486, 2002.

[8] Z. Haas and T. Small. A new networking model for biological applications of ad hoc sensor networks. *IEEE/ACM Trans. on Netw.*, 14(1):27–40, 2006.

[9] T. Karagiannis, J.-Y. Le Boudec, and M. Vojnovic and. Power law and exponential decay of intercontact times between mobile devices. *IEEE Trans. Mobile Comput.*, 9(10):1377 –1390, 2010.

[10] I. Rhee, M. Shin, S. Hong, K. Lee, S. Kim, and S. Chong. On the levy-walk nature of human mobility. *IEEE/ACM Trans. on Netw.*, 19(3):630–643, 2011.

[11] M. Rytgaard. Estimation in the Pareto distribution. *ASTIN Bulletin*, 20(2):201 – 216, 1990.

[12] M. Seshadri, S. Machiraju, A. Sridharan, J. Bolot, C. Faloutsos, and J. Leskove. Mobile call graphs: beyond power-law and lognormal distributions. In *ACM SIGKDD'08*, pages 596–604. ACM, 2008.

[13] T. Spyropoulos, K. Psounis, and C. Raghavendra. Efficient routing in intermittently connected mobile networks: The multiple-copy case. *IEEE/ACM Trans. on Netw.*, 16(1):77–90, 2008.

[14] A. Vahdat and D. Becker. Epidemic routing for partially connected ad hoc networks. Technical report, 2000.

# Information Dissemination Dynamics in Delay Tolerant Network: A Bipartite Network Approach

Sudipta Saha[*]
Indian Institute of Technology
Kharagpur-721302, India
sudipta.saha
@cse.iitkgp.ernet.in

Niloy Ganguly[†]
Indian Institute of Technology
Kharagpur-721302, India
niloy
@cse.iitkgp.ernet.in

Animesh Mukherjee[‡]
Indian Institute of Technology
Kharagpur-721302, India
animeshm
@cse.iitkgp.ernet.in

## ABSTRACT

In this paper, we present a model of a delay tolerant network (DTN) and identify that this model can be suitably reformulated as a bipartite network and that the major predictions from the former are equivalent to that of the latter. In particular, we show that the *coverage* of the information dissemination process in the DTN matches accurately with the size of the largest component in the suitably thresholded one-mode projection of the corresponding bipartite network. In the process of this analysis, some of the important insights gained are that (a) arbitrarily increasing the number of agents participating in the dissemination process cannot increase the *coverage* once the system has reached the stationary state for a given buffer time (i.e., the time for which a message resides in the buffer of the places visited by the agents), (b) the *coverage* varies inversely with the square of the number of places in the system and directly with the square of the average social participation of the agents and (c) it is possible to design an optimal buffer time for a desired cost of *coverage*. To the best of our knowledge, this is the first such work that employs the rich theoretical backbone of bipartite networks as a "proxy" for the analysis of the otherwise intractable DTN dynamics thus allowing for various novel analytical estimates.

[*]PhD Student, Department of Computer Science & Engineering, Indian Institute of Technology, Kharagpur, India.
[†]Associate Professor, Department of Computer Science & Engineering, Indian Institute of Technology, Kharagpur, India.
[‡]Assistant Professor, Department of Computer Science & Engineering, Indian Institute of Technology, Kharagpur, India.

## Categories and Subject Descriptors

C.2.1 [**Computer-communication Networks**]: Network Architecture and Design—*Store and forward networks, Wireless communication*; G.2.2 [**Discrete Mathematics**]: Graph Theory—*Network Problems*

## General Terms

Performance,Theory

## Keywords

Delay Tolerant Network, DTN, Bipartite Network, Information Dissemination, Coverage

## 1. INTRODUCTION

The medium of communication among the devices in a DTN is wireless and therefore the connectivity among them is of short range [1] and inefficient. This creates a lack of certainty in direct communication among the devices. Therefore, the concept of indirect communication is introduced in DTN which involves storing a message temporarily in a certain 'throwbox' [26] or 'buffer' in different places so that other devices can pick up the message when the agents carrying the devices, visit these places. Consequently, it is reasonable to advocate that the performance of any search or information dissemination application, developed for such a network, is strongly influenced by the mobility pattern of these participating agents. The destination of the mobile agents in general are selected on a purely random basis as is the case for the random way point model [5] and various other similar models [4, 6]. However, in a realistic scenario, it becomes important to incorporate the social behavior of the agents (as in for humans) that has been recently introduced for instance *'Self Similar Least Action Walk'* based mobility model [15]. A very important component of such observation is that, the agents have a tendency to visit places depending on the attractiveness of those places. In other words, there is a preferential choice driving the mobility pattern of the agents. The preferential choice induces long and short range correlation among the walkers. This increases the complexity in the mobility pattern making it extremely difficult to use traditional mathematical techniques, like mean field theory, in calculating the *coverage*, i.e., the number of distinct nodes receiving the message in

[1]The radius of communication of the devices using technology such as bluetooth [7], is of the order of a few meters.

the steady-state. In fact, this is possibly one of the most important reasons why such an analysis is almost inexistent in the literature although there have been many works related to the design of dissemination algorithms directed to maximize coverage in both wired and wireless networks [2, 10, 17, 18].

In this paper, we identify that, the message dissemination process in DTN has a natural and one-to-one correspondence with a time varying bipartite network where one partition contains a fixed number of places and the other partition contains the agents whose number continuously grows starting from zero. Hence, we try to show that, without "re-inventing the wheel" (such as recasting the DTN scenario as an epidemic spreading process or otherwise), the existing rich theoretical backbone of the evolving bipartite network [8, 16, 19] can be exploited to analyze the otherwise intractable characteristics of the message dissemination process in DTN. In principle, we concentrate on the analysis of the coverage in terms of the largest component of the *one-mode projection* of the underlying bipartite network suitably thresholded by a time-varying threshold.

In the following, we first present a brief survey of the works already done to model different aspects of DTN. In section 3 we describe a realistic scenario of information dissemination in DTN and subsequently examine it under the lens of the analytical framework of the bipartite networks (section 4). In section 5, first we show that the time evolution of the fraction of nodes to which a message gets disseminated (i.e., coverage) in the DTN has a perfect overlap with the growth of the largest component size of the one-mode projection (explained later) of the bipartite network suitably thresholded by a time varying threshold. Next, we show the correlation of the parameters in the two domains. Finally, we provide a closed form expression for the largest component size of the thresholded one-mode projection which in turn gives the theoretical estimate for the coverage achievable in the DTN.

## 2. RELATED WORKS

Routing/dissemination of information in DTN have been in focus for a long period of time. Many algorithms have been developed to solve these problems. For example, in the store-carry-forward paradigm based algorithms, the mobility of the agents is exploited to convey message packets. In these strategies, the devices carried by the agents temporarily buffer the data and forward it to other (appropriate) agents. Epidemic routing [11], spray and wait protocols [24] are some examples of these. Analysis and subsequent use of contact history among the agents have also been the focus of various works [3, 14, 25].

Modeling DTN through different analytical framework has also gained much interest in recent times. Epidemic modeling [20], ordinary differential equations [23], partial differential equations [1] or Markov models [21] have been successfully used to represent DTN. Performance of different routing strategies have been evaluated using these mathematical techniques. The primary objectives of these studies were to analyze the data delivery ratio and data delivery latency.

Augmenting the DTN with different types of stationary message storing devices, such as throwbox [26], has been the recent trend to enhance the communication opportunities in between the mobile devices. In [22], the authors show that, use of such relay devices effectively decreases the data delivery latency as well as increases the data delivery ratio. However, the existing modeling or analysis related works on DTN do not consider the presence of such message buffers. Recently, Gu et al [13] have addressed this issue in similar lines as of those presented here. They have proposed the use of message buffers as an instance of bio-inspired methods (e.g., pheromone or footprint). Using discrete Markov chain based modeling, they analyzed the importance of buffer time, i.e., the amount of time a message copy can stay in a message buffer, as well as the preferences of visiting different places by the mobile agents. They studied the impact of these two crucial system features on the latency and the message delivery ratio of the dissemination process in the network. In our work we specially emphasize on the inherent bipartite nature of the "buffer augmented DTN" and thus bring forward the fact that instead of starting from scratch, the existing theories of the bipartite network can be used (with necessary modification) to analyze the coverage problems related to DTN.

## 3. INFORMATION DISSEMINATION IN DTN

We consider a certain number of mobile agents ($t$) who participate in the information dissemination process and a certain number of common places ($N$) where the agents usually go. An agent is assumed to make $\mu$ number of visits to different places (hence, $\mu$ directly models the social participation of the agents). The place to be visited next by an agent, is chosen preferentially from the pool of places where the preference to be given on a place is directly proportional to the number of other agents who already visited the place. A sequential agent arrival pattern is assumed which implies that the next agent will join the system after the previous agent has visited all of the $\mu$ places. To make the process realistic, we introduce a concept of *time* which denotes the count of the agents who have joined the system and visited all the places they were supposed to visit. Within a single *time* unit an agent creates $\mu$ number of connections ($\tau$ varies from 1 to $\mu$). To follow the store and forward paradigm, we assume that, each of the places as well as the agents has a buffer ('throwbox' [26]) where several pieces of information can be stored. Throughout this paper we assume that all the communications between the agents take place via these message buffers. Without any loss of generality, we consider here the dissemination of a single message. Due to the limited size, a message will be discarded from the buffer of a place, after a certain *time* duration $b$, termed as buffer time [2]. However, due to sequential agent arrival pattern, we assume that the agents can store a particular message for their full life time, after they have got the message. The information dissemination process along with the observable in the process are described below.

- *Initial Condition*: Initially, i.e., prior to the start of

---

[2] The buffer time happens to be a very crucial factor in the message dissemination process. A longer buffer time implies plenty of redundant message copies, higher CPU cycle as well as higher battery power consumption in the mobile devices. On the other hand small buffer time may imply insufficient node coverage. Therefore, arriving at an optimal buffer time is crucial for the system to maintain a balance between performance and load. Consequently, this factor plays as one of the most significant constraints in the system.

the dissemination process, the buffers of each place and device of the agents are assumed to have enough space to participate in the dissemination process.

- *Start*: The initiator of the dissemination process brings or creates a message in its buffer and sequentially visits $\mu$ number of places (preferentially) where the messages are dropped so that other agents can pick up the message and participate in the dissemination process.

- *Dissemination through places*: When some agent comes in a place where the initiator has already dropped the message, the message gets transferred to the buffer of the agent if the agent is not already containing the message.

- *Dissemination through other agents*: Once some agent picks up a message from some place, it also participates in the dissemination process. When such an agent visits a place where the message is not present, the message gets transferred to the buffer of the place. At the same time, the buffer timer of the place gets set to the value $b$ which implies that the message can be stored in the place for $b$ *time* units. This provides some chances to the place to convey the message to different agents.

- *Buffer timer manipulation*: After the end of each *time* unit, the buffer timers in all the places decrease by one. However, when an agent who has already got the message from some place, arrives at another place where the message is already there, we assume that a fresh copy of the message is brought to the place. We relate this event to the importance of the message which is being disseminated and therefore give advantage to the place by setting its buffer timer value back to the maximum value, i.e., $b$.

- *Observable*: Keeping the 'on the fly' [21] structure of DTN in mind, we abstract out the definition of coverage as number of different places a particular message can reach, under a given message dissemination scheme. Hence, in the process described above, we measure, the number of distinct places, the buffers of which contain the message at different *time* steps. We denote this quantity by $G_d$. This specific quantity is of interest from the perspective of dissemination because, the probability that any other mobile agent will receive the information while visiting a place, is directly proportional to the value of $G_d$.

Figure 1 pictorially describes this information dissemination process. In the next section we describe the analysis of this process of dissemination using evolution of the bipartite network.

## 4. MODELING BY BIPARTITE NETWORK

In this section, we describe the modeling of the whole information dissemination dynamics in DTN as a bipartite network with one growing partition. We visualize the DTN dynamics as a bipartite network where one of the partitions corresponds to the places while the other corresponds to the agents. The number of places is fixed and finite ($=N$) while the number of agents grows over *time* and is modeled by

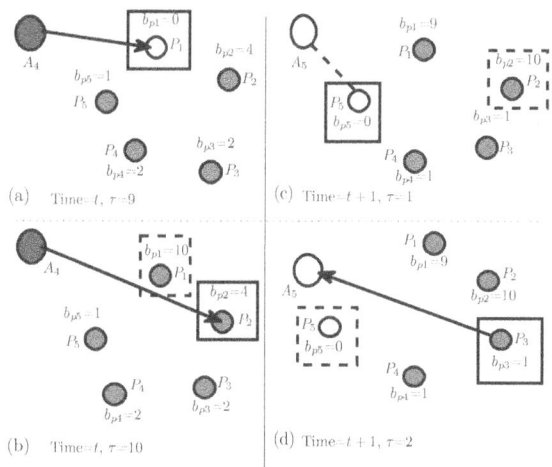

Figure 1: A schematic diagram showing the information dissemination process running in a DTN. There are 5 places: $P_1$ ... $P_5$ and 2 agents $A_1$ and $A_2$. The notation $b_{pi}$ denotes the current value of the buffer timer of $P_i$, i.e., the rest *time* units for which the message can be stored in the buffer of place $P_i$. We assume here that the maximum buffer *time* is 10($=b$) and the agents can travel 10($=\mu$) places sequentially. The empty and filled up circles denote the absence and presence of message respectively. Arrows denote the direction of transfer of the message. Part (a) and (b) are snaps of the process at the beginning of creating the $9^{th}$ and $10^{th}$ connection at *time* $t$ by agent $A_4$. In (a), the message gets transferred to $P_1$ from $A_4$. In (b), the buffer timer of $P_2$ gets set to 10. Part (c) and (d) are snaps at the beginning of creating $1^{st}$ and $2^{nd}$ connection by the agent $A_5$. In (c), no message transfer happens. In (d), the message gets transferred from $P_3$ to $A_5$. Most of the possible interactions are shown through this figure.

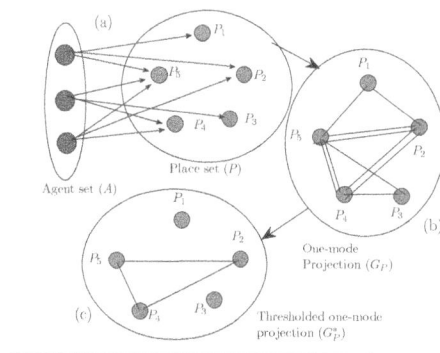

Figure 2: Schematic diagram of a possible scenario of the bipartite network corresponding to a DTN comprising five places, i.e., $N$=5 ($P_1...P_5$) and three agents, i.e., $t$=3 ($A_1$ ... $A_3$) with $\mu$=3. The diagram shows a possible status after all the agents have joined the system. Part (a) shows the bipartite network, (b) is the one-mode projection and (c) is the thresholded one-mode projection for threshold value 2.

the parameter $t$. Each agent is allowed to make $\mu$ connections sequentially one by one, each time choosing a place in a preferential fashion (see Figure 2). Therefore, in both the bipartite network as well as the DTN domain, the parameters $N$, $\mu$ and $t$ have the same significance. Table 1 summarizes the precise relationships between the parameters in the two domains.

## 4.1 One-mode projection in bipartite network

In the course of visiting from place to place, the underlying store and forward paradigm based algorithm, operating in the mobile devices of the agents, allows one agent to convey a message from one place to other. In order to capture this message flow, we take the one-mode projection of this bipartite network on the place set (one-mode projection, on the place set, is a place to place graph where two places are connected by an edge if there is one common agent who has visited/connected both of the places). In this projection, we assign weights to the edges where a particular weight denotes the number of parallel edges between the two places via same or different agents (if one agent has visited two places each twice, then there will be four parallel edges in the one-mode projection which captures the fact that there are actually four different possible communications between the two places). For a bipartite network $G$, we denote its one-mode projection on the place set by $G_P$ (see Figure 2(b)).

## 4.2 Thresholding edge weight in the bipartite network

It can be intuitively understood that, the buffer time in the information dissemination process in a DTN, actually controls the flow of the message from one place node to the other. Therefore, the probability that a common visit will convey a message, is directly proportional to the buffer time $b$. Hence, for a given value of $b$, there is a minimum number of common visits required to effectively convey a message from one place to the other. To reflect this scenario corresponding to a buffer time $b$ in DTN, we introduce a threshold edge weight $c$ in the bipartite network. In particular, we prune those edges in the one-mode projection of the bipartite network whose edge weights fall below $c$. Hence, the rest graph contains only those edges which represent strong and stable inter-place communication and thereby accurately simulates the effect of $b$ in DTN. For a certain bipartite network $G$ we denote this thresholded one-mode projection on the place set, by $G_P^*$ (see Figure 2(c)).

## 4.3 Time-varying threshold

The weights of the edges in $G_P$ vary with time for continuous arrival of the agents and their connection patterns. Therefore, in order to bring the concept of temporal stability of the edge weights between places of the bipartite network (always imposed by $b$ in DTN over the entire time evolution of the system), we calculate the threshold $c$ as a function of *time* $t$, i.e., number of agents who already joined the system (as we consider the arrival rate as one agent per time step). In this work, we assume that $c \propto t$ and hence $c = v \times t$. We take this constant of proportionality $v$ as the characteristic parameter equivalent to a buffer time $b$ in DTN.

Table 1: Relationship between the parameters in DTN and the bipartite network

| Type | DTN | Bipartite Network | Remarks |
|---|---|---|---|
| Parameters | Agents($t$) | Agent partition ($t$) | Growing |
| | Places($N$) | Place partition ($N$) | Fixed and finite |
| | Number of place an agent visits ($\mu$) | Number of connections an agent creates with different places ($\mu$) | Constant (can be taken from some specified distribution also) |
| | Buffer time ($b$) | Threshold varying with $t$ ($v$) | (See subsection 4.3) |
| Observable | Number of places where the message could reach under the dissemination process ($G_d$) | Size of the largest component of the thresholded one-mode projection ($G_b$) | These quantities should match |

## 5. RESULTS

The coverage in DTN for a certain value of $b$ is conceptually the same as the size of the largest component (denoted by $G_b$) in the one-mode projection of the bipartite network suitably thresholded (keeping other parameters same in the two domains). We simulate the time evolution for both $G_d$ as well as $G_b$ for many different parameter combinations. We find that these quantities become independent of $t$ (become stable) after a certain number of agents have joined the system and their time evolutions also match accurately with each other for different value pairs of $v$ and $b$. Figure 3 shows few such sample cases for different combinations of the other two parameters $N$ and $\mu$.

Thus, it can be understood that, to analyze the coverage, i.e., number of distinct sites where a message can be spread by means of a specific information dissemination scheme in DTN, the very first step to be carried out is to estimate the parameters of the bipartite network setup that can work as a perfect "proxy" for the DTN setup. Therefore, this parameter estimation process is our next focus.

## 5.1 Estimation of parameters of bipartite network from DTN

In order to successfully establish a correlation between the two domains we need to find a value of the time varying threshold $v$ in bipartite network functionally equivalent to the value of the parameter $b$ in DTN. In other words, the relationship between these two parameters creates a bridge between the two domains. To understand this equivalence we start from the ground state in the DTN, where there is no buffer i.e. $b=0$. In that case the information will not pass on to a single other node. Therefore, the coverage will be zero. This situation is captured in the bipartite network by using a sufficiently high value of $v$ which prunes all the parallel edges in the one-mode projection of the bipartite network resulting in a set of isolated nodes only. Similarly, in the opposite case,

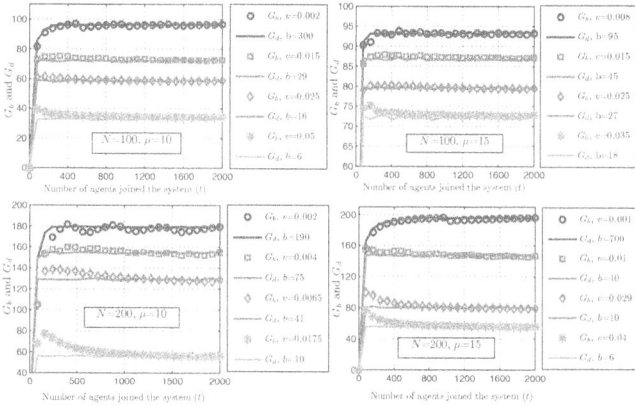

Figure 3: Comparison of the size of the largest component ($G_b$) in the thresholded one-mode projection of the bipartite network and the number of places in DTN where the information, being disseminated, is found i.e. *coverage* ($G_d$). The bipartite network is formed corresponding to a DTN set up consisting of ($N$=) 100 common places and ($t$=) 2000 mobile agents each of which creates ($\mu$=) 10 connections with the places i.e. visits 10 places sequentially. The four parts of the figure shows the results for four different combinations of $N$ and $\mu$.

Figure 4: Plots in (a) and (b) show the relationship of $G_d$ and $G_b$ with the buffer time $b$ and the time varying threshold $v$ respectively for different values of $\mu$ while the value of $N$ is set at 100. Plots in (d) and (e) show the same relationship for different values of $N$ while the value of $\mu$ is set at 15. Plots in (c) and (f) show the relationship between $b$ and $v$ for different values of $\mu$ and $N$ along with the parameters $A$, $C$ and $\alpha$ for the best fitting instance (99% confidence level) of the equation 1. In all these cases the values of $G_b$ and $G_d$ have been collected after enough number of users have joined the system and the system has stabilized.

if we employ a sufficiently high value of $b$, eventually almost all the nodes in the DTN will receive the message (after sufficiently large number of agents have joined the system with a sufficiently high value of $\mu$). This scenario can be mimicked by a very small value of $v$ ($\approx 0$) in the bipartite network which does not prune any of the edges from the one-mode projection. Hence, an inverse relationship among these two parameters can be observed. It can also be realized that the relationship between $v$ and $b$, is not independent of $N$ and $\mu$. We extensively simulate the relationship between $v$ and $b$ for different values of $N$ and $\mu$. Figure 4 shows few sample results and the nature of the increase in the $G_d$ and decrease in the $G_b$ with the increase in $b$ and $v$ respectively for few combinations of $N$ and $\mu$. Using the commonality of $G_d$ and $G_b$, we infer values of $v$ for given values of $b$. We find that all these relationships between $v$ and $b$ (for different combinations of $N$ and $\mu$) fit the following equation-

$$v = Ab^{-\alpha} + C \qquad (1)$$

where $A$ and $C$ are certain constants. Figure 4 also shows the values of the parameters $A$, $C$ and $\alpha$ for few different combinations of $N$ and $\mu$. Through extensive simulation we find that the value of the exponent $\alpha$ is directly proportional to a non-linear combination of $N$ and $\mu$ that we plan to explore further as a part of our future work.

Next we focus on the theoretical analysis of the size of the largest component in the thresholded one-mode projection of the bipartite network for a given value of $v$. We describe this process in the following subsections.

## 5.2 Component formation in bipartite network

In real life, the selection process for the next place to be visited by the mobile agents, incorporates a little ran-

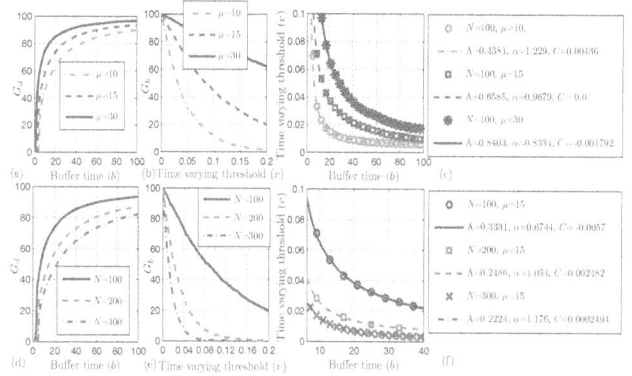

domness, rather than being fully preferential. However, for simplicity purpose, in this work, we have assumed a pure preferential selection model. Due to this reason, the place to place graph generated after application of threshold on the one-mode projection of the bipartite network on the place set, exhibits a special property described below. We denote this property by $\mathbb{P}$.

$\mathbb{P}$: *After any number of agents have joined, the thresholded one-mode projection of the bipartite network on the place set, consists of a single connected component while the rest of the places that are not part of the largest component are degenerate, i.e., have degree zero.*

The full proof of this property (as empirically observed by us for the first time in Figure 5) is out of the scope of this paper. However, the basic intuition behind this scenario can be sketched as follows. Due to preferential attachment, the nodes of the connected component necessarily have high degree in the bipartite network. Hence, any new agent would almost surely make some of the connections with these nodes. Conversely, none of the new agents will make all its connections only with the isolated nodes. Hence, isolated nodes will either get absorbed in the giant component or stay as single entity.

In order to test the above hypothesis, we simulate the evolution process of the bipartite network for various combinations of the parameters $N$, $\mu$ and $v$ for large values of $t$. For pure preferential attachment process, we always find that the sum of the two quantities : *size of the largest component*, i.e., $G_b$ and *the number of components* (denoted by $C_b$), is equal to $N+1$. Plots of Figure 5 show the evolution of $G_b$ and $C_b$ for two different values of $v$ in four different combinations of $N$ and $\mu$. It is clear from the plots that the relationship $G_b+C_b=N+1$, is maintained through-

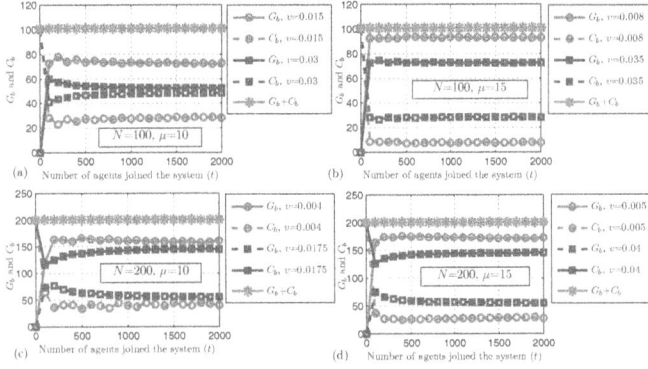

Figure 5: Comparison of the number of components ($C_b$) and largest component size ($G_b$) in four different combinations of $N$ and $\mu$. For each such combination, the relationship is shown for two sample values of $v$ throughout the evolution of the bipartite network as 2000(=t) agents join the system sequentially. Each plot also shows the sum of the two quantities (i.e. $C_b + G_b$) in all these cases.

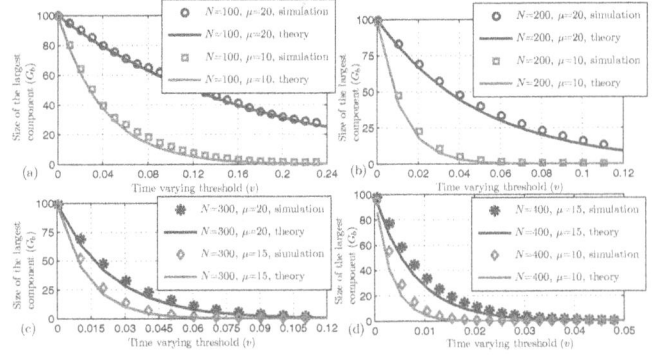

Figure 6: Comparison of the size of the largest component in the bipartite network obtained from equation 3 and the same obtained from simulation of the evolution process in the bipartite network under different combinations of $N$ and $\mu$.

out the whole evolution of the bipartite network for all $v$ which implies that the property $\mathbb{P}$ is satisfied throughout the process. This result would help us to calculate the size of the largest component.

## 5.3 Calculation of the size of the largest component

In this subsection, we use the existing theory of bipartite network to derive the expected coverage in the described information dissemination process in DTN. We denote the ground state of the bipartite network $G$ as $G_0$ where set $A$ is empty and set $P$ contains $N$ places. One agent joins the set $A$ per time step and creates $\mu$ connections with some elements in the set $P$. We consider here only the fully preferential attachment process. Using the basics of Polya Urn scheme and the de Finetti theorem [9], it can be shown that the probability that a new agent will create a connection with a place $i$ in $P$ (let us denote this probability as $\theta_i$) of $G$, is marginally Beta distributed with the parameters $b_i$ and $b_0$ where $b_i$ is the initial degree of node $i$ in $G_0$ and $b_0$ is the sum of the degrees of the other nodes in $P$ of $G_0$. The work in [12], assumes that all the nodes in set $P$ have same initial degree 1, i.e., $b_i=1$, $\forall i \in P$ which implies that $\theta_i$, $\forall i \in P$, are identically Beta distributed with parameters $(1, N\text{-}1)$. It has been shown in [12] that the expected number of edges at large time between node $i$ and $j$ of set $P$ is $(\mu^2 - \mu)\theta_i\theta_j$. From this it has been shown that at large time, the probability that a place $i$ with attractiveness $\theta_i$ will be connected to some other place in the thresholded one-mode projection of $G$, i.e., $G_P^*$ (conversely the probability that a place $i$ with attractiveness $\theta_i$ have more than $c = v \times t$ parallel edges with some other node in the one-mode projection, i.e., $G_P$) is equal to $\left(1 - \frac{v}{(\mu^2-\mu)\theta_i}\right)^{(N-1)}$. Finally, deriving the expected number of such connections of node $i$ and using the Beta distribution of the attractiveness of the places of set $P$, [12] develops the cumulative degree distribution of the nodes

of the set $P$ in $G_P^*$. Equation 2 shows this cumulative degree distribution which effectively gives the probability that a randomly selected node in $G_P^*$ has degree greater or equal to $k$ at large time $t$.

$$F_k(t) = \left(1 - \frac{v}{(\mu^2 - \mu)x}\right)^{(N-1)} \qquad (2)$$

where $v$ is the time varying threshold and $x = 1 - \left(\frac{k}{N-1}\right)^{\frac{1}{N-1}}$.

We use this result for calculating the size of the largest component as follows. As we consider only fully preferential model of attachment, the property $\mathbb{P}$ holds true throughout the evolution process. Hence, the fraction of nodes which form the largest component are the nodes that have degree 1 or higher. This fraction could be obtained by putting $k=1$ in equation 2. We multiply this probability with $N$ to get the number of nodes in the largest component which reads as follows.

$$G_b = N \times \left[1 - \left(\frac{\sqrt[N-1]{(N-1)}}{\sqrt[N-1]{(N-1)} - 1}\right) \times \left(\frac{v}{\mu^2 - \mu}\right)\right]^{N-1} \qquad (3)$$

We simulate the evolution of the bipartite network for various combinations of $N$ and $\mu$ and measure the size of the largest component. Figure 6 shows the match of theory and the simulation results for eight such different combinations.

For large value of $N$, the ratio $\left(\frac{\sqrt[N-1]{(N-1)}}{\sqrt[N-1]{(N-1)} - 1}\right)$ is almost equal to 1. The value $v$ is generally less than 1 and $(\mu^2 - \mu)$ is comparatively a large value ($\gg 1$). Hence, ignoring the higher order terms in binomial expansion in equation 3, we get the following simplified form of $G_b$.

$$G_b = N - \frac{N(N-1)}{\mu(\mu-1)} \times v \qquad (4)$$

## 5.4 Calculation of the coverage in DTN

Putting the exact expression for $v$ in terms of $b$ in equation 4 we get the following formula which provides a very close estimation of the coverage in DTN.

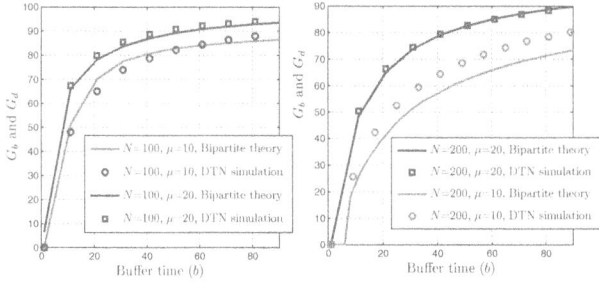

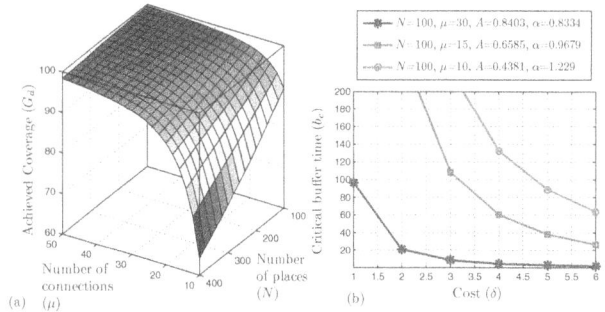

Figure 7: Comparison of the size of the largest component, i.e., $G_b$ in the bipartite network obtained from equation 3 and the *coverage* values obtained by simulating the information dissemination process in DTN for various buffer times under four different combinations of $N$ and $\mu$.

Figure 8: Part (a) shows the plot of the percentage of the number of places covered (i.e., $G_d$) as a function of the total number of places ($N$) and the number of places an agent visits ($\mu$). Part (b) shows the plot of the critical buffer time value ($b_c$) constrained by different cost values ($\delta$).

$$G_d(= G_b) = N - \frac{N(N-1)}{\mu(\mu-1)} \times (Ab^{-\alpha} + C) \qquad (5)$$

To test the accuracy, we simulate the dissemination process in DTN, for several buffer times (for different combinations of $N$ and $\mu$) for which we already know the relationship with their equivalent time varying threshold. Using that relationship we calculate the equivalent values of $v$ and also calculate the size of the largest component in the thresholded one-mode projection of the equivalent bipartite network from equation 3. We find that these theoretical results and the coverage obtained by simulation of information dissemination in DTN, match quite accurately thus pointing to the fact that the empirically derived relationship between $v$ and $b$ is quite appropriate. Figure 7 shows four such cases for different combinations of $N$ and $\mu$.

## 5.5 Insights obtained from the analysis

We arrive at several significant insights from the closed form expression of coverage derived in terms of $N$, $\mu$ and $b$ (equation 5). Most importantly, it has been found that, for a given value of $N$, $\mu$ and $b$, the number of nodes covered in the dissemination process does not grow unboundedly with the increase in the number of agents ($t$) joining the system; rather, after a certain value of $t$, the total number of place nodes covered, gets stabilized and is limited by the buffer time $b$ (see Figure 3).

Further, to visualize the significance of equation 5, we present a three-dimensional plot (Figure 8(a)) showing the interaction of $G_d$, $\mu$ and $N$. It is clear from the plot that $G_d$ bears an inverse relationship with $N$ and a direct relationship with $\mu$. Precisely, a closer look at the equation 5 reveals that indeed $G_d$ is inversely proportional to $N^2$ and directly to $\mu^2$.

As a final remark, we note that the rate of growth of $G_d$ slows down with an increasing value of $b$ (see Figure 7) and particularly after a critical value (say $b_c$) $G_d$ is almost stable. From the perspective of a design engineer, the value of $b_c$ could be crucial. Naturally the cost associated with dissemination process (buffer space, CPU consumption, bandwidth etc.), should bear a connection with $b_c$. We assume that this cost is a ratio of the coverage achieved to the overhead in-

curred per unit increase of $b$ and denote it by $\delta$. Therefore, the rate at which the cost increases should not overshoot the rate at which $G_d$ increases and so we have the following relationship.

$$\frac{dG_d}{db} > \delta \qquad (6)$$

Evaluating the derivative of $G_d$ with respect to $b$ and using relationship 6, we get the following expression for the critical buffer time ($b_c$).

$$b_c = \sqrt[\alpha+1]{\frac{\alpha A N(N-1)}{\delta \mu(\mu-1)}} \qquad (7)$$

In Figure 8(b), we plot the values of $b_c$ for different values of $\delta$ under few different combinations of $N$ and $\mu$. The value of $\delta$ can be chosen freely according to the design requirement. However as a proof of concept we inspect the values of delta between 1 and 6 to investigate the nature of the equation 7. The values of $b_c$, for say $N=100$ and $\mu=10$, effectively mean that, to satisfy different values of $\delta$, the employed values of $b$ should be below the corresponding curve in Figure 8(b). It is seen that, with increasing cost, increasing the buffer time rapidly becomes uneconomical.

## 6. CONCLUSION

In this paper, we have identified a novel way of looking at the problem of estimating the coverage of the information dissemination process in delay tolerant networks. In particular, we found that the DTN system has an underlying "bipartite mechanism" which can be used as a "proxy" for analytically estimating the coverage. We have shown that the complexity of computing this quantity for DTN can be reduced to the problem of inferring a suitable value of $v$ (bipartite domain) from a given value of $b$ (DTN domain). In addition, we also observed that (i) arbitrarily increasing $t$ does not amount to an arbitrary increase in the coverage while constrained by a specific value of $b$, (ii) the coverage achievable is inversely proportional to $N^2$ and directly proportional to $\mu^2$ and (iii) it is possible to design an optimal value of $b$ for a desired cost of coverage.

Some of the limitations of the current approach are that

we assumed a sequential arrival of agents and a fully preferential choice of their movements. It is quite straightforward to relax both of these assumptions by respectively allowing for overlapping life span of the agents in both the domains and introducing a randomness parameter in the model that can control the preference factor of the agents. Preliminary experiments on both of these issues (to be reported elsewhere), show that the major trends are (almost) equivalent to what has been presented in this paper. Furthermore, in this work we mainly focus on the coverage achieved after the system has stabilized/saturated. However, the coverage achieved within a given period of time is also of high importance. Many works have been already done to analyze this in general peer-to-peer networks [17, 18]. We plan to incorporate this factor as a part of the future extension of this work.

# 7. ACKNOWLEDGMENTS

This work was funded by a project under *Department of Information and Technology* (DIT), Govt. of India. The authors are grateful to Saptarshi Ghosh for his comments and suggestions that helped in improving the quality of the paper.

# 8. REFERENCES

[1] J. Y. L. B. A. Chaintreau and N. Ristanovit'c. The age of gossip: Spatial mean-field regime. In *Proceedings of ACM Sigmetrics*, 2009.

[2] N. Alon, C. Avin, M. Koucky, G. Kozma, Z. Lotker, and M. R. Tuttle. Many random walks are faster than one. In *Proceedings of SPAA*, pages 119–128, 2008.

[3] A. Balasubramanian, B. Levine, and A. Venkataramani. Dtn routing as a resource allocation problem. In *SIGCOMM Computer Communication Review*, 2007.

[4] C. Bettstetter. Mobility modeling in wireless networks: categorization, smooth movement, and border effects. *Sigmobile Mob. Comput. Commun. Rev.*, 5:55–66, July 2001.

[5] J. Broch, D. A. Maltz, D. B. Johnson, Y. C. Hu, and J. Jetcheva. A performance comparison of multi-hop wireless ad hoc network routing protocols. In *Proceedings of MobiCom*, pages 85–97, 1998.

[6] T. Camp, J. Boleng, and V. Davies. A survey of mobility models for ad hoc network research. *WCMC: Special Issue*, 2:483–502, 2002.

[7] O. S. Chau, P. Hui, and V. O. K. Li. An architecture enabling bluetooth[tm]/jini[tm] interoperability. In *Proceedings of PIMRC*, pages 3013–3018, 2004.

[8] M. Choudhury, N. Ganguly, A. Maiti, A. Mukherjee, L. Brusch, A. Deutsch, and F. Peruani. Modeling discrete combinatorial systems as alphabetic bipartite networks: Theory and applications. *Phys. Rev. E*, 81:036103, 2010.

[9] P. Diaconis and D. Freedman. Finite exchangeable sequences. *Annals of Probability*, 8(4):744–764, 1980.

[10] V. V. Dimakopoulos and E. Pitoura. On the performance of flooding-based resource discovery. *IEEE Trans. Parallel Distrib. Syst.*, 17(11):1242–1252, 2006.

[11] W. Gao, Q. Li, B. Zhao, and G. Cao. Multicasting in delay tolerant networks: A social network perspective. In *Proceedings of MobiHoc*, 2009.

[12] S. Ghosh, A. Srivastava, T. Krueger, and N. Ganguly. Degree distribution of one mode projection of evolving alpha bipartite networks. *Available at http://cse.iitkgp .ac.in/resgrp/cnerg/Files/BipartiteNetwork.pdf.*

[13] B. Gu, X. Hong, and P. Wang. Analysis for bio-inspired thrown-box assisted message dissemination in delay tolerant networks. *Telecommun Syst, Springer*, 2011.

[14] R. Jathar and A. Gupta. Probabilistic routing using contact sequencing in delay tolerant networks. In *Proceedings of COMSNETS*, 2010.

[15] K. Lee, S. Hong, S. J. Kim, I. Rhee, and S. Chong. Slaw: A new mobility model for human walks. In *Proceedings of INFOCOM*, pages 855–863, 2009.

[16] A. Mukherjee, M. Choudhury, and N. Ganguly. Understanding how both the partitions of a bipartite network affect its one-mode projection. *Physica A*, 390(20):3602 – 3607, 2011.

[17] S. Nandi, L. Brusch, A. Deutsch, and N. Ganguly. Coverage-maximization in networks under resource constraints. *Phys. Rev. E*, 81(6):061124, Jun 2010.

[18] K. Oikonomou, D. Kogias, and I. Stavrakakis. A study of information dissemination under multiple random walkers and replication mechanisms. In *Proceedings of MobiOpp*, pages 118–125, 2010.

[19] F. Peruani, M. Choudhury, A. Mukherjee, and N. Ganguly. Emergence of a non-scaling degree distribution in bipartite networks: a numerical and analytical study. *Euro. Phys. Lett.*, 79:28001, 2007.

[20] F. Peruani, A. Maiti, S. Sadhu, H. Chatt'e, R. R. Choudhury, and N. Ganguly. Modeling broadcasting using omnidirectional and directional antenna in delay tolerant networks as an epidemic dynamics. *IEEE-JSAC*, 28(4):524, 2010.

[21] A. Picu and T. Spyropoulos. Distributed stochastic optimization in opportunistic networks: The case of optimal relay selection. In *Proceedings of CHANTS*, 2010.

[22] S. Shahbazi, S. Karunasekera, and A. Harwood. Improving performance in delay/disruption tolerant networks through passive relay points. In *Wireless Network*, 2011.

[23] T. Small and Z. J. Haas. The shared wireless infostation model: a new ad hoc networking paradigm (or where there is a whale, there is a way). In *Proceedings of MobiHoc*, 2003.

[24] T. Spyropoulos, K. Psounis, and C. S. Raghavendra. Spray and focus: Efficient mobility-assisted routing for heterogeneous and correlated mobility. In *Proceedings of IEEE PerCom workshops*, 2007.

[25] Q. Yuan, I. Cardei, and J. Wu. Predict and relay: an efficient routing in disruption-tolerant networks. In *Proceedings of MobiHoc*, 2009.

[26] W. Zhao, Y. Chen, M. Ammar, M. Corner, B. Levine, and E. Zegura. Capacity enhancement using throwboxes in dtns. In *Proceedings of MASS*, pages 31 –40, 2006.

# Utility-based Forwarding: a Comparison in Different Mobility Scenarios

Elena Pagani
Computer Science Dept., Università degli Studi
di Milano, Italy
CNR-IIT, National Research Council, Italy
pagani@dico.unimi.it

Gian Paolo Rossi
Computer Science Department
Università degli Studi di Milano, Milano, Italy
rossi@dico.unimi.it

## ABSTRACT

Several proposals are available in the literature that deal with the problem of message forwarding in Opportunistic Networks (ONs). These proposals attempt to derive the path from source to destination that minimizes delivery latency and traveled hops, and maximizes the probability of successful delivery, while saving the overall system resources through a limitation of the number of message copies. Utility-based forwarding achieves these goals through the use of functions that discriminate among nodes in terms of their *utility* to reach a destination. Although the approach is very promising, so far, there is no understanding about the tight relationship between utility functions and the mobility scenario in which they operate and, as a consequence, we are unable to design efficient solutions for practical ONs.

In this work, we focus on this point by analysing five well known utility functions in five different scenarios. We establish relationships between the mechanisms adopted by the utility functions to discriminate among candidate relays, and the characteristics of the environment in terms of people mobility and the structure of their communities. The results can be useful to select an appropriate forwarding mechanism when deploying an experimental Opportunistic Network, and to design a novel utility function able to adapt to variable mobility patterns.

## Categories and Subject Descriptors

C.2 [**Computer-Communication Networks**]: Network Architecture and Design

## Keywords

Opportunistic networks, Utility-based forwarding

## 1. INTRODUCTION

The recent advances in short range radio technologies are driving the search for new wireless networking platforms that complement the 3G/4G network infrastructure with the aim of offloading cellular networks and providing a flexible alternative to deliver emerging mobile computing services such as, those location-sensitive (i.e targeted advertising, recommending systems) and contact-sensitive (i.e. mobile social networking, content sharing, urban sensing) [23, 12]. In this emerging heterogeneous networking scenario, Opportunistic Networks (ONs) [20] have great potential as a viable solution to enable the communications between the content source (either a mobile user in the neighborhood or a roadside AP) and the target mobile user(s). This communication is obtained by deploying a multi-hop path on top of contacts amongst mobile devices. The practical feasibility and the efficiency of such a path mainly depend on the function of forwarding.

The research has longly studied forwarding in ONs and proposed a variety of solutions in the literature [1, 2, 7, 8, 9, 10, 14, 16, 18]. For evident motivations, the most viable and practical solutions are those following the single-copy approach or a controlled multi-copy. The challenge is to approximate the performance and efficiency of the unviable, but optimal, forwarding achieved by using an oracle knowing all future contacts. Several interesting and practical solutions are available that attempt to approximate the oracle decisions by assigning utility values to nodes on the base of the available set of past and current contacts. If the values are properly assigned the forwarding is more likely able to discriminate, among the set of contacts, the relays that belong to the optimal path towards destination. The bet is that the past behavior of a node will be maintained in the future. In order to effectively discriminate among relays of different quality, contact dynamics should show two relevant behaviors. First, the contact pattern *within* every pair of nodes needs to be somehow regular. Irregular encounters make impossible to forecast the node likelihood of encountering the destination. Secondly, contact dynamics *between* different pairs of nodes need to be somehow heterogeneous, otherwise any node is an equally good candidate for data relaying.

Social and human sciences tell us that the human attitude of grouping in social communities, of sharing common locations or of commuting from one location to another, actually generates the required regularities and heterogeneities [21, 19, 22, 6]. This makes the approach promising but concentrates all challenges on the choice of the utility function. A good utility function captures the proper behavior of a given mobility pattern and is thus able to discriminate between good relay nodes (with more chances to encounter the destination in the future) and the others. A bad utility function is unable to do this and considers all nodes as equally useful.

For the above arguments, it is easy to argue that utility-based forwarding is highly influenced by the nature of underlying mobility patterns. Despite that, nobody has analyzed utility functions in different mobility scenarios. As a consequence, today we are unable to select the utility function that better fits with the nature of a given practical, deployable networking scenario or to design a good utility function for a given mobility setting. This paper focuses on these still open points by analyzing 5 well known utility functions in 5 different mobility scenarios. The results of the paper should be useful when deploying an experimental ON or to design a novel utility able to adapt to changing mobility patterns.

This is not the only paper approaching this argument; in [9], a subset of the utility functions we consider here have been analyzed; however, the analysis covers a very short time interval (3 hours) and uses real traces reproducing densely populated environments only.

## 2. MOBILITY SCENARIOS

ONs are supposed to be active in the mobile periphery of the Internet, the belt centered around the locations where users live, work or socialize. We can assume that each location covers a limited area (let us say at most $1000 \times 1000$ m.), involves at most a few hundred people and is characterized by mobility patterns that reflect the type of sociality people have there. In fact, the social destination of a place influences the way communities are formed, how people interact and move from group to group etc., thus conditioning the behavior of regularities and heterogeneities. With the intention of describing relevant human mobility and social attitudes we devised 5 scenarios that are described in the following.

The first scenario describes human attitudes in workplaces. Individuals in such a scenario are highly sedentary, but, due to the limited space, sooner or later also individuals belonging to separate communities (let us call them unfamiliar) happen to encounter one another. The mobility setting is characterized by contacts among familiar people lasting for long time and by unfrequent contacts between unfamiliar nodes. By slightly varying the previous scenario we can represents two new conditions that reproduce the human attitudes in spaces with higher mobility. In the former, very closed communities are considered. Sporadically, an individual belonging to a community may temporarily visit a different community. By contrast, a few individuals (namely, the *travelers*) continuously go back and forth thus having frequent contacts with people belonging to different communities. In the second, the community boundaries are softened and people belonging to different communities have more chances to encounter one another. In both cases, the contact duration is shorter than before – due to higher mobility – and the amount of global contact opportunities is higher. Two extreme conditions are also considered: a random and a deterministic mobility scenario. The latter yields perfectly predictable contacts; similar conditions can be found, for instance, in mobility patterns of a public transportation system.

### 2.1 Implementation of the Scenarios

We reproduce the five scenarios above by means of, respectively, a real trace, a synthetic model (HCMM [3]), and

two benchmark models (Random Waypoint and a perfectly deterministic encounter pattern). We adopted HCMM because it can model human mobility patterns according to the influence of both popular locations and social relations. This is in contrast with, for instance, [13, 5, 17] that only consider one of the two aspects. Moreover, the setting of the HCMM parameters is very simple, easier than, for instance, in [15].

*PMTR traces.*

The real traces are obtained through a campus experiment and are generated by 44 people equipped with wireless devices, named PMTRs, with 10 m. radio range [11].[1] In order to compare the trace from a real dataset with those obtained from synthetic models, we eliminated nights and weekends from the experimental dataset, thus producing a dataset covering 13 working days, from 8:00 AM to 8:00 PM (156 h.). These samples are more than the 80% of the samples obtained from the whole experiment.

*HCMM model.*

In the two scenarios produced with HCMM, 44 nodes move in a $1000 \times 1000$ m. area with speed in [0.5, 1.5] m/s for 156 hours; the transmission range is 10 m. As an initial interaction matrix, we used weights derived from the number of contacts between pairs of nodes in the PMTR trace. The highest number of contacts was assigned weight 0.9. The weight associated to half of the average number of contacts has been adopted as a threshold to derive the connection matrix. No reconfiguration is performed and the remaining probability is set to 0.8. In the scenario named *HCMM_det5*, the next cell is chosen deterministically and the rewiring probability is 0.1; we adopted 5 travelers. In the *HCMM_pro1* scenario, the probabilistic criterion is adopted, with rewiring probability of 0.3 and 1 traveler.

*Random Waypoint.*

We produced a RWP scenario with BonnMotion v1.5 [4], with 44 nodes moving for 156 hours over a $500 \times 500$ m. area, with speed varying in the interval [0.5, 1.5] m/s. We adopted a pause interval of 3600 s. The trace produced by BonnMotion is post-processed in order to derive the contacts between nodes, assuming that the communication range is 10 m. and with sampling granularity of 1 s.

*Deterministic model.*

A deterministic (DET) scenario has been produced in the same environment as RWP as follows: communities of a 10% of nodes each have been set. Then, for each pair of nodes $i$ and $j$, contacts are generated with parameters drawn from the mean values obtained with PMTRs (Table 1). The first contact occurs at a random time within the first hour. All contacts occur with a fixed inter-contact time ICT $c_{ij}$ from one another and have a fixed contact length $l_{ij}$. In our experiments, $c_{ij}$ has been extracted with uniform distribution within the range [1000, 1500] s. for nodes in the same community, and [5000, 6000] s. otherwise, while $l_{ij}$ always varies

---

[1]The data set can be downloaded from the CRAWDAD archive (http://www.crawdad.org/unimi/pmtr).

**Table 1: Comparison between synthetic scenarios and PMTR traces**

| scenario | # contacts | length (s.) | ICT (s.) | $E(\sigma(\text{ICT}_{ij}))$ | $\sigma(E(\text{ICT}_{ij}))$ | #hops $\mathcal{O}$ | latency $\mathcal{O}$ | %dest $\mathcal{O}$ |
|---|---|---|---|---|---|---|---|---|
| DET | 146.83 | 452.12 | 3369 | 0 | 1250 | 3.41 | 323 s. | 100 |
| PMTR | 15.59 | 521.98 | 3381 | 3709 | 4198 | 3.22 | 32207 s. | 87.8 |
| HCMM_det5 | 292.28 | 11.88 | 1599 | 14738 | 21922 | 4.08 | 4367 s. | 87.3 |
| HCMM_pro1 | 58.75 | 11.94 | 8814 | 17243 | 20527 | 3.69 | 1273 s. | 100 |
| RWP_3600 | 22.05 | 123.03 | 24149 | 23674 | 6127 | 3.54 | 2130 s. | 99.7 |

in the range [300, 600] s. This will allow to differentiate nodes in terms of their utility to reach a certain destination.

In both DET and RWP, all pairs of nodes eventually have a contact; this is not true in the other settings. However, in DET, nodes in the same group are more used to encounter one another, i.e. they encounter with a higher rate than with other nodes. In HCMM_det5, communities are quite closed, and their members mainly communicate through travelers. In PMTR, communities are not that closed, but people are sedentary and encounters occur seldom and are quite long. HCMM_pro1 still produces communities – unlikely RWP – but they are often mixed. As far as HCMM traces are concerned, we verified that the ICTs they produce fit with a Pareto's distribution. In Table 1, we show a comparison among the characteristics of the PMTR traces and those obtained by synthetic mobility models. All indices are averaged over all pairs of nodes and concern the whole time window. For PMTR, the ICT is evaluated only for consecutive contacts occurring within the same day. In columns 5 and 6 of Table 1, we report respectively: (i) the average on all pairs of nodes of the standard deviation of the ICT between node i and node j; (ii) the standard deviation among all pairs of nodes of the average ICT between node i and node j. The former gives an indication about the regularity of contacts *within* every pair. The latter gives an indication of the heterogeneity *between* different pairs of nodes. In Table 1, the performance achieved with an oracle-based optimal forwarding $\mathcal{O}$ is also reported, as the number of hops, latency, and percentage of reached destinations from each node to every other.

## 3. UTILITY-BASED FORWARDING

We study utility functions that capture various aspects of human interactions by using different mechanisms. In this work, we do not consider approaches involving community detection (e.g. [14]). Indeed, community detection is computationally expensive and still difficult to implement in a distributed manner. We do not assume that knowledge is a-priori available about movements, as in [18, 1]. We rather focus on mechanisms allowing to gain knowledge on contact dynamics in order to infer future encounters, e.g. through the analysis of the encounter history, as in [16, 9, 8], or the estimation of the centrality of nodes, as in [7]. This knowledge can be used for unicast communications. In particular, the following functions are considered:

- **Greedy (G) online:** the utility of a relay increases with the number of times it has encountered the message destination so far [9].

- **Greedy-total (GT) online:** the utility of a relay increases with the number of encounters (with any node) it has observed so far [9].

**Table 2: Comparison between utility functions**

| | **G** | **GT** | **F** | **P** | **SB** |
|---|---|---|---|---|---|
| coloc. vs. social | coloc | coloc | coloc | coloc | social |
| dest. dependent | Yes | No | Yes | Yes | mix |
| aging | No | No | Yes | Yes | No |
| transitivity | No | No | No | Yes | No |

- **Fresh (F):** a node $n1$ has greater utility than $n2$ for a destination $D$ if the last encounter of $n1$ with D occurred more recently than that of $n2$ [8]. This approach has no memory of the past.

- **Prophet (P):** in [16], three mechanisms are used to maintain utilities. When two nodes $n1$ and $n2$ encounter, each one increments its delivery probability to the other as follows: $P(n1, n2) \leftarrow P(n1, n2) + (1 - P(n1, n2)) \cdot P_{init}$. A transitivity property of encounters is considered, such that, if $n2$ encountered $n3$, then the delivery probability of $n1$ to $n3$ is $P(n1, n3) \leftarrow P(n1, n3) + (1 - P(n1, n3)) \cdot P(n1, n2) \cdot P(n2, n3) \cdot \beta$. To take into account changing contact dynamics, an aging function is applied before each utility exchange such that $P(n1, n2) \leftarrow P(n1, n2) \cdot \gamma^t$, where $t$ is the time elapsed since the last update.

- **SimBet (SB):** the utility of a node $n1$ for a destination $D$ depends on both $n1$'s betweenness (i.e., whether it belongs to the shortest path between any two other nodes), and its similarity with $D$ (i.e., whether there are nodes encountered by both $n1$ and $D$). We estimate both indexes as in [7].

In Table 2, we report a comparison among the characteristics of the considered approaches in terms of: whether colocation or social aspects are considered, whether the approach is destination dependent or not, whether it includes aging and transitivity mechanisms.

### 3.1 Measuring the Utility

In order to measure the goodness of a generic utility function $\mathcal{U}$, we consider its distance from optimal routing, able to follow the shortest path between source and destination.

To capture the dynamics of the utility values, we proceed as follows: every $M$ minutes we freeze the utilities held at that time by nodes. According to those values we compute the ($\mathcal{U}$) path followed from each source to each destination, and we compute the distance between such path and the optimal one. Two indices $\Delta_e$ and $\Delta_l$ are considered to evaluate the distance and obtained as follows:

$$\Delta_e = \frac{\#hops \ of \ \mathcal{U} \ path \ - \ \#hops \ of \ \mathcal{O} \ path}{\#hops \ of \ \mathcal{O} \ path}$$

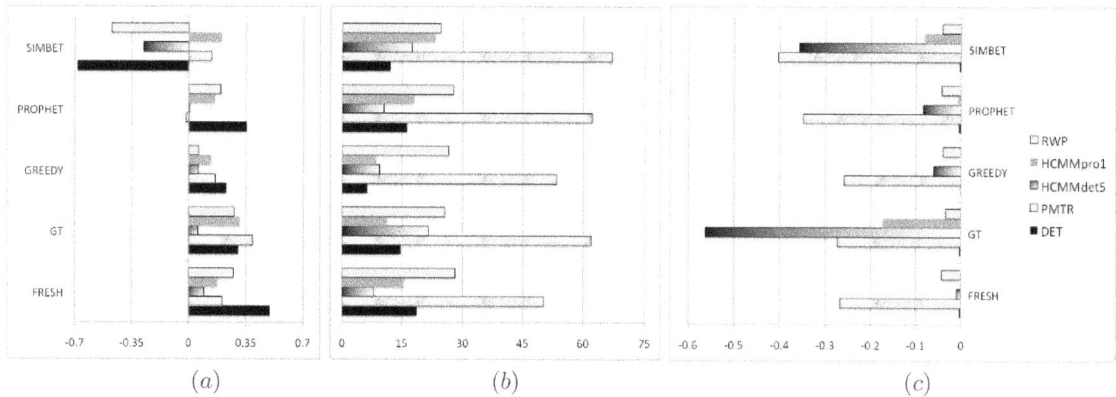

Figure 1: Comparison among approaches in terms of $(a)$ $\Delta_e$, $(b)$ $\Delta_l$, and $(c)$ $\Delta_d$.

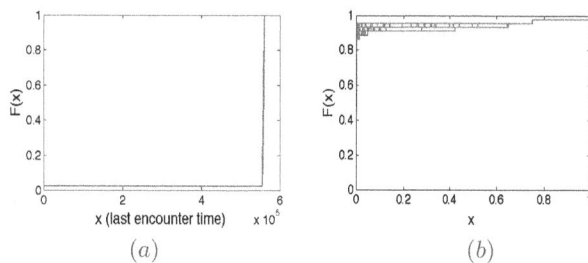

Figure 2: DET: *ecdf* of the utility values (one plot for each node), for $(a)$ Fresh, and $(b)$ Prophet.

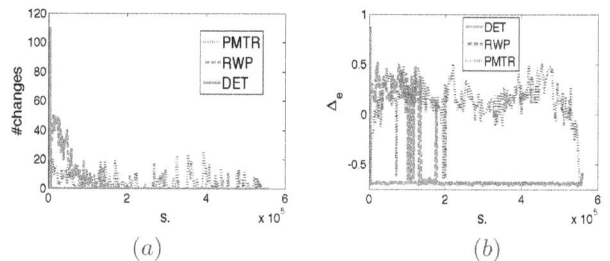

Figure 3: SimBet: $(a)$ utility changes, and $(b)$ $\Delta_e$ vs. time.

$$\Delta_l = \frac{latency\ of\ \mathcal{U}\ path\ -\ latency\ of\ \mathcal{O}\ path}{latency\ of\ \mathcal{O}\ path}.$$

We use the path length in hops as an indirect measure of the overall system energy consumption (the higher the hop count, the more the devices involved to forward third party traffic over a radio channel). $\Delta_l$ measures how worse is the service offered to the users. $\Delta_e$ and $\Delta_l$ are measured just for the destinations reached by both $\mathcal{O}$ and $\mathcal{U}$. An index $\Delta_d$ is computed similarly, accounting for the fraction of destinations that $\mathcal{U}$ reaches with respect to $\mathcal{O}$. Clearly, if $\mathcal{U}$ follows paths with the same characteristics as $\mathcal{O}$, its distances will be 0.

## 4. SIMULATIONS

We used simulations to observe performance and behavior of the described utility functions in the different mobility scenarios. The following general settings have been adopted. The sampling rate is $M = 30$ minutes. This is also the length of the warm up period before generating the first message (this lets the utilities to initialize). The sampling is stopped after $(7 \times 12)$ hours leaving the last $(6 \times 12)$ hours for message delivery. Results are averaged over all source-destination pairs. The same single-copy forwarding policy is used for all approaches: if $n1$ has a message $m$ for $D$ and it encounters $n2$ with greater utility for $D$ than its own, then $n1$ forwards $m$ to $n2$ and removes $m$ from its buffer.

Some more specific settings follow. Prophet simulations use the following parameters [16]: $P_{init} = 0.75$, $\beta = 0.25$, $\gamma = 0.98$. In SimBet, the parameter $\alpha$, weighing between similarity and betweenness, is set to $\alpha = 0.5$ [7].

## 4.1 Results

In this Section, we focus on the simulation results with the aim of understanding the impact of different mobility settings on the performance of utility functions, and of capturing what mechanisms – among those the functions adopt – are successful to accurately discriminate among the utility of candidate relays. For a given mobility model, a good utility function is able to assign high values to nodes with more chances of encountering the destination in the future, while it assigns lower values to the other nodes. By contrast, a utility function that happens to flatten the assigned values cannot behave properly. Fig.1 summarizes the performance of the different approaches with the considered mobility traces. We now analyze in detail the relevant situations for each mobility model.

Let us consider the DET model first. The challenge in DET is identifying the groups of nodes that are used to encounter one another with the higher frequency, and assigning them a high utility. Here most of the approaches do not work well for slightly different motivations. All of them can hardly discriminate between nodes belonging to the community of the destination (that more likely meet the destination) and the other nodes (as shown by the *ecdf* of the utility values for the various destinations (Fig.2)). In Fresh, a node outside the community of the destination $d$ might be chosen as forwarder because it encountered $d$ very recently although it has few opportunities of being co-located with $d$. The performance of Prophet is negatively affected by the aging mechanism that flattens all the utilities to small values. As a consequence, the utility of nodes in the same community (*familiars*) is slightly higher than those of outside nodes (*un-*

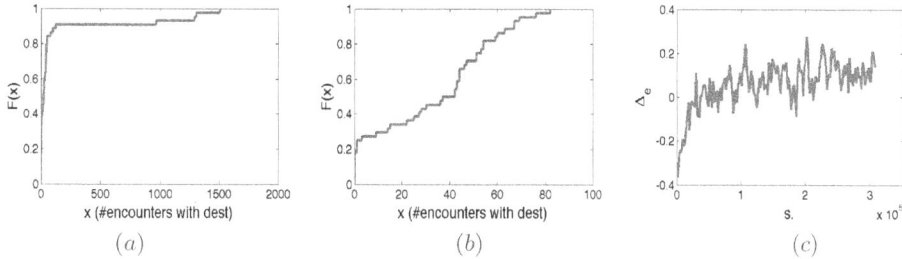

Figure 5: **Greedy in HCMM_det5:** *ecdf*(utilities) for (*a*) a sedentary node, and (*b*) a traveler. (*c*) $\Delta_e$ vs. time.

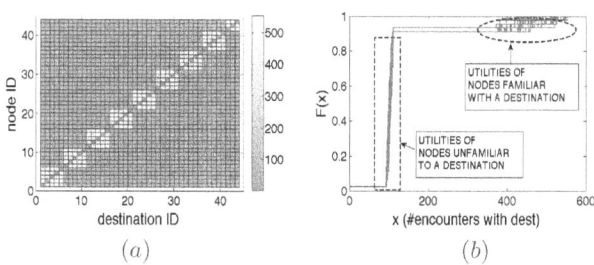

Figure 4: **Greedy in DET:** (*a*) **utilities, and** (*b*) *ecdf* **of the utility values (one plot for each node).**

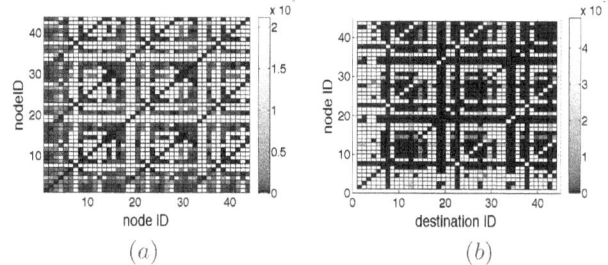

Figure 6: (*a*) **ICTs with HCMM_det5, and** (*b*) **utilities learnt by Fresh.**

*familiars*), but both are extremely low (0.107 and 0.018 on average respectively). Under these conditions, at each forwarding a very small progress towards the destination may be obtained, thus negatively affecting both latency and number of hops. A similar behavior is shown by Greedy Total: as the pattern of encounters is really similar for all nodes, utilities for the nodes are aligned and no node clearly distinguishes as preferred forwarder. An almost random choice does not guarantee that the destination is approached.

SimBet is more qualitative than quantitative: for each pair of nodes it records whether they have encountered or not. In DET (but similar arguments apply to RWP), when all nodes have encountered one another at least once, then no node is between any source-destination pair, and every node is equally similar to each other (both have encountered all the other nodes). At this point, utilities are all equal, and the approach is forced to behave as a *direct contact*, as there is no node better than the source to reach the destination. In order to show this, we measured the frequency of changes of the utilities associated to the nodes. To measure this index, as before we measure the number of changes in the utilities occurring in every time window of M minutes. The frequency drops to 0 after at most 6000 s. (the longest ICT) with DET, and a couple of days with RWP (Fig.3(*a*)).[2] When this occurs, all nodes are equal, and $\Delta_e$ becomes negative indicating that forwarding is through direct contact(Figg.1(*a*) and 3(*b*)).

By contrast, Greedy remembers the whole past history and properly detects the familiarity of a node with the destination. Thus, it perfectly catches the existence of 9 communities in DET (Fig.4(*a*)): for every node there is a sharp difference between the utilities of nodes outside and inside

the destination community (Fig.4(*b*)). Hence, as soon as a node in the same community of the destination is encountered, it is adopted as a relay, and its utility guarantees that further forwarding can only occur within the destination community. After a short learning phase of around 2-3 hours, Greedy is able to characterize paths that reach all the destinations with performance comparable to that of $\mathcal{O}$.

In HCMM_det5, the challenge should be to capture the few travelers. Greedy utilities discriminate among travelers, nodes resident in a community different from that of a destination $d$, and nodes familiar with the destination. A sedentary node unfamiliar with $d$ has utility for $d$ equal to 0 (Fig.5(*a*), left part). By contrast, a node familiar with $d$ encounters it many times and its counter is high (Fig.5(*a*), right part). Travelers show the behavior in Fig.5(*b*).[3] Travelers move around and encounter almost all nodes, but for few times: notice the difference in the $x$-axis between Fig.5(*a*) and (*b*). Fig.5(*b*) also shows that travelers do not visit all communities with uniform probability. The consequence on forwarding is that: (*i*) a node unfamiliar with $d$ forwards the message to either a traveler or a node familiar with $d$ that it happens to encounter; (*ii*) a traveler forwards the message to a node familiar with $d$. When a traveler is adopted as relay, its utility guarantees that further forwarding can only involve either a traveler more accustomed to visit the destination's group, or the destination's group itself. As soon as the destination's group is reached, forwarding is confined within it. Fig.5(*c*) shows that initially – during the first 8 hours roughly – counters (utilities) are not yet well differentiated, and direct contact is often used.

The behavior of Fresh and Prophet in HCMM_det5 can

---

[2]For the sake of comparison, the behavior with one setting where not all nodes encounter is shown. But, for readability, not all models are shown.

[3]Plots in Fig.5(*a*) and (*b*) are for two specific nodes, but their behavior is common to all residents and travelers respectively.

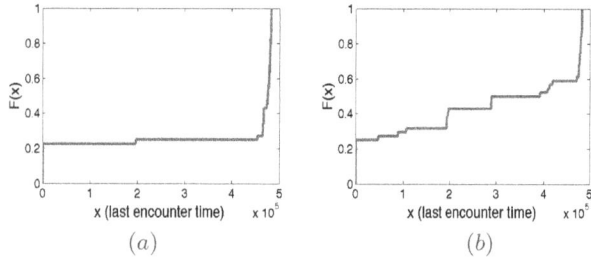

Figure 7: Fresh in HCMM_det5: *ecdf*(utilities) for (*a*) a sedentary node, and (*b*) a traveler.

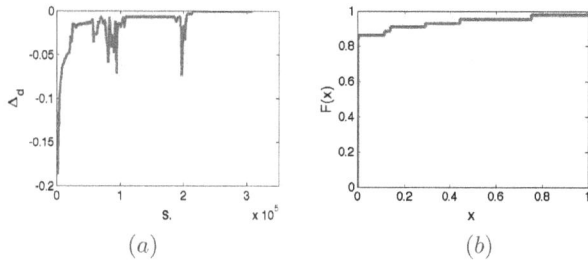

Figure 8: HCMM_det5 – (*a*) Fresh: $\Delta_d$ vs. time. (*b*) Prophet: *ecdf*(utilities) for a node.

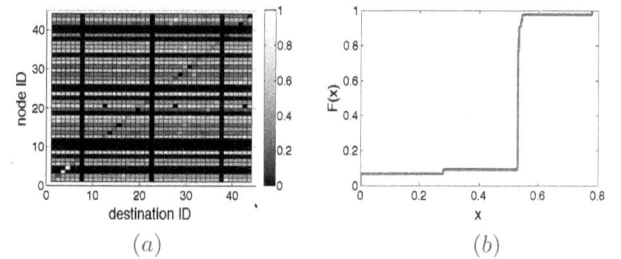

Figure 9: SimBet in HCMM_det5: (*a*) utilities for every pair of nodes, and (*b*) *ecdf*(utilities) for a certain node.

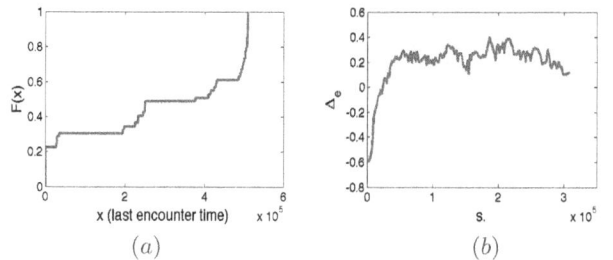

Figure 10: Fresh in PMTR: (*a*) *ecdf*(utilities) for a node. (*b*) $\Delta_e$ vs. time.

be explained with similar arguments. Although the mobility pattern is close to DET, both the approaches perform better. In fact, despite Fresh has no memory, the habit of frequent encounters within groups continuously refreshes the time of the last encounter with a destination and lets emerge the familiarity.[4] In Fig.6, the value of the ICT's and the Fresh utilities are shown for all pairs of nodes, in (*a*) and (*b*) respectively. Sedentary nodes do not see unfamiliar nodes (Fig.7(*a*), left side), while have very recent encounters with nodes in their group (Fig.7(*a*), right side). By contrast, travelers have some nodes visited very recently (Fig.7(*b*), right side) – possibly of the last visited community – nodes visited in a more or less recent past (Fig.7(*b*), middle), and nodes never encountered (Fig.7(*b*), left side). The very same classification of nodes is achieved as with Greedy, with similar results. The successful learning of familiarities is brought into evidence by the variation of the reached destinations along time: after a learning period of around 7-8 hours, where some losses are experimented, all destinations are reached as with $\mathcal{O}$ (Fig.8(*a*)).

As far as Prophet is concerned, the aging mechanism virtually resets utilities deriving from sporadic visits to unfamiliar communities. As in Fresh, utilities of familiar nodes are continually refreshed. By contrast, the utility of travelers derives from two contrasting forces: on the one hand, the aging decreases the utility for nodes belonging to communities visited long ago. On the other hand, the transitivity mechanism increases the utility of travelers for nodes not encountered, but belonging to most visited communities (i.e., whose familiars have been encountered by the traveler). As a results, the utility values are slightly better differentiated (Fig.8(*b*)), and again acceptable performance is achieved.

In this environment, the approaches unable to single out useful relays are those partly or completely destination-inde-

---

[4]Notice the small average ICT in Table 1.

pendent, namely, Greedy Total and SimBet. With the former, a node continuously encountering its familiars – but never visiting unfamiliar groups – could have a high counter. With SimBet, each node has very homogeneous utilities for all destinations, indicating that the destination-independent component of the utility function is predominant (Fig.9(*a*)). Similarly, the *ecdf*'s over the utilities show only a big step (an example is shown in Fig.9(*b*)) distinguishing between the nodes never encountered and all the others. The nodes with high utilities are more than the travelers; they could also be nodes connecting two communities just because they belong to one and happened to visit the other. Such a node could be chosen as forwarder, but it gives no guarantees of encountering a destination resident in a different group.

The PMTR setting is quite similar to HCMM_det5, although the higher sedentariness and longer contact duration impose much higher latencies than in the other settings (Fig.1(*b*)). Yet, these characteristics help Fresh utilities in differentiating among familiar users (recently encountered), groups frequented less often, and users never seen (the multiple steps in Fig.10(*a*)). As in HCMM_det5, Fresh is able to learn these differences within roughly 12 hours (Fig.10(*b*)), and they are continuously refreshed thanks to users' habits in spite of the Fresh lack of memory. Noticeably, the differences among different degrees of familiarity are reported more sharply by Fresh utilities than by Greedy utilities (Fig. 11). We conjecture that this is due to the fact that the area is relatively small, and almost all pairs of nodes are likely to encounter soon or late. In such an environment, familiarity could be better caught by considering the duration of the encounters, rather than their number as Greedy does. For all nodes, the *ecdf* over Greedy utilities shows a behavior similar to that of travelers in HCMM_det5 (Fig.5(*b*)).

The other approaches are penalized by the characteristics of the environment. Greedy Total differentiates very well

34

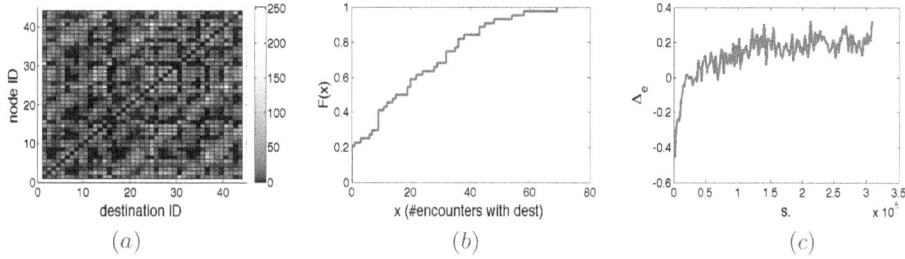

Figure 13: Greedy in HCMM_pro1: (a) utilities for every pair of nodes; (b) *ecdf*(utilities) for a node. (c) $\Delta_c$.

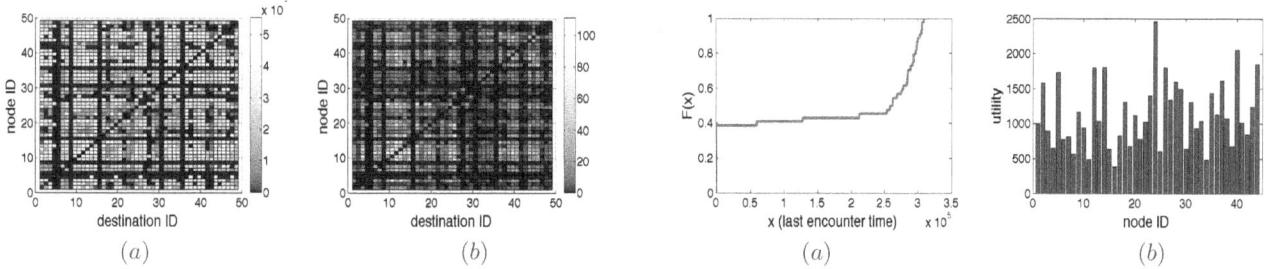

Figure 11: PMTR: utilities for every pair of nodes for (a) Fresh, and (b) Greedy.

Figure 14: HCMM_pro1 – (a) Fresh: *ecdf*(utilities) for a node. (b) Greedy Total: utilities of nodes.

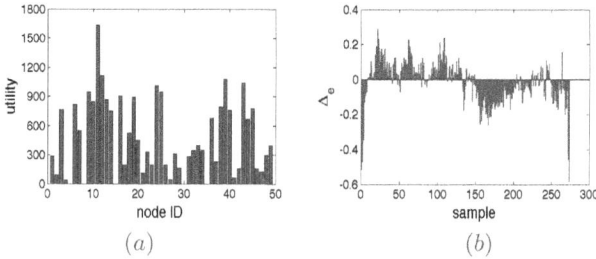

Figure 12: PMTR – (a) Greedy Total: utilities. (b) Prophet: $\Delta_e$ vs. time.

among the popularity of different people (Fig.12(a)). We verified that nodes with high utilities correspond to individuals either with many cooperations, or whose office is in a central position where people frequently pass by. As before, utilities of SimBet mainly depend on betweenness. Though, as with HCMM_det5, destination-independence leads to inappropriate choice of the forwarders – which are not guaranteed to encounter the destination in the near future – and thus to poor performance. Due to high sedentariness, the utilities of Prophet are drastically reduced by the aging mechanism, making difficult for social structures to emerge, and thus hindering the identification of appropriate forwarders. As a consequence, direct contact is often used for message delivery (Fig.12(b)).

In HCMM_pro1, the accumulation of the whole history performed by Greedy makes it able to differentiate the communities, although mixed, as it emerges by both the utilities reported in Fig.13(a) (they are not uniform, as instead it happens in RWP) and the behavior of the *ecdf* over utilities (which shows more encounters with a subset of nodes) (Fig.13(b)). As a consequence, Greedy is able to discover the communities (although with a long learning phase: roughly

24 hours) and to move from a direct contact policy to an adequate choice of forwarders, as shown in Fig.13(c)

The mixing of communities in HCMM_pro1 negatively affects Fresh ability of distinguishing different degrees of familiarity with a destination. Utilities either show nodes never encountered, or nodes recently encountered, which are all the others due to the mix (Fig.14(a)). Similarly, Prophet utilities do not make nodes habits to emerge. Mixing however helps the two approaches in reaching all the destinations, although with longer latencies than Greedy. By contrast, the destination-oblivious approach taken by both Greedy Total and SimBet, although able to differentiate among nodes with different popularity (Fig.14(b)), in spite of mixing communities, chooses forwarders that do not guarantee of eventually reaching the destinations, as highlighted by Fig.1(c).

*In summary.*

From the above considerations we can say that:
• in all environments, the maintenance of the whole history allows to accurately discriminate among relays;
• the approaches that either do not maintain or forget the past show good discrimination capability when the people habits continuously refresh the information about recurrent encounters, and there is a sharp difference between the encounter dynamics of familiar and unfamiliar nodes;
• as a consequence of the above argument, the approaches that maintain a qualitative record of the past encounters do not let recurrences to emerge, thus flattening the differences among relays;
• destination independency does not allow to identify the relays that are more likely approaching a given destination; this makes packets traveling away from the optimal paths.

## 5. CONCLUSIONS

In this paper, the behavior of five utility functions for message forwarding in ONs is studied, in five different mobility models. The results allow to characterize what mechanisms are able to discriminate among the usefulness of various relays depending on the environment characteristics. Hence, indications emerge that can drive the choice of the appropriate policy for computing the utilities according on the people mobility and the confinement of their communities.

Several developments are possible. Modifications of the mechanisms can be studied, in order to overcome difficulties that they may face in some environments. We are designing a utility mechanism able to adapt to the mobility characteristics of the people involved in an ON. As a future work, we plan to adapt some of the considered approaches in order they are able to "follow" several nodes pooled by a common interest, instead of a unicast destination.

## 6. ACKNOWLEDGMENT

The authors thank Chiara Boldrini for supplying the HCMM code. This work was funded partially by the Italian Ministry for Instruction, University and Research under the PRIN PEOPLENET (2009BZM837) Project, and partially by the European Commission under the FP7 SCAMPI (258414) Project.

## 7. REFERENCES

[1] S. Ahmed, S.S. Kanhere. HubCode: Message Forwarding using Hub-based Network Coding in Delay Tolerant Networks. In *Proc. MSWiM*, 2009, pp. 288-296.

[2] C. Boldrini, M. Conti, J. Jacopini, A. Passarella. HiBOp: a History Based Routing Protocol for Opportunistic Networks. In *Proc. WOWMOM 2007*, pp. 1-12.

[3] C. Boldrini, M. Conti, A. Passarella. Users Mobility Models for Opportunistic Networks: the Role of Physical Locations. In *Proc. IEEE WRECOM*, 2007, pp.1-6.

[4] –. BonnMotion: A mobility scenario generation and analysis tool. University of Bonn, Institute of Computer Science 4, Communication and Networked Systems. http://net.cs.uni-bonn.de/wg/cs/applications/bonnmotion/

[5] V. Borrel, F. Legendre, M. De Amorim, S. Fdida. SIMPS: Using Sociology for Personal Mobility. *IEEE/ACM Transactions on Networking*, 17(3), 2009, pp. 831-842.

[6] A. Chaintreau, P. Hui, J. Crowcroft, C. Diot, R. Gass, J. Scott. Impact of Human Mobility on the Design of Opportunistic Forwarding Algorithms. *Proc. IEEE INFOCOM 2006*, pp. 1-13.

[7] E. Daly, M. Haahr. Social Network Analysis for Routing in Disconnected Delay-Tolerant MANETs. In *Proc. MobiHoc 2007*, pp. 32-40.

[8] H. Dubois-Ferriere, M. Grossglauser, M. Vetterli. Age matters: efficient route discovery in mobile ad hoc networks using encounter ages. In *Proc. ACM MobiHoc 2003*, pp. 257-266.

[9] V. Erramilli, A. Chaintreau, M. Crovella, C. Diot. Diversity of forwarding paths in pocket switched networks. In *Proc. ACM/SIGCOMM IMC* (Oct. 2007), pp. 161-174.

[10] V. Erramilli, M. Crovella, A. Chaintreau, C. Diot. Delegation Forwarding. In *Proc. MobiHoc 2008*, pp. 251-259.

[11] S. Gaito, E. Pagani, G.P. Rossi. Strangers help friends to communicate in opportunistic networks. *Computer Network* 55(2), Elsevier, February 2011, 374-385.

[12] B. Han, P. Hui, V.S.A. Kumar, M.V. Marathe, G. Pei, A. Srinivasan. Cellular Traffic Offloading through Opportunistic Communications: A Case Study. *Proc. CHANTS 2010*.

[13] W. Hsu, T. Spyropoulos, K. Psounis, A. Helmy. Modeling Time-Variant User Mobility in Wireless Mobile Networks. In *Proc. IEEE INFOCOM 2007*, pp.758-766.

[14] P. Hui, J. Crowcroft, E. Yoneki. Bubble rap: social-based forwarding in delay tolerant networks. In *Proc. 9th ACM Int. Symp. on Mobile Ad Hoc Networking and Computing*, 2008.

[15] K. Lee, S. Hong, S. J. Kim, I. Rhee, S. Chong. SLAW: A Mobility Model for Human Walks. In *Proc. IEEE INFOCOM 2009*.

[16] A. Lindgren, A. Doria, O. Schelñen. Probabilistic Routing in Intermittently Connected Networks. In *Proc. 1st International Workshop on Service Assurance with Partial and Intermittent Resources*, 2004, pp. 239-254.

[17] A. Mei, J. Stefa. SWIM: a Simple Model to Generate Small Mobile Worlds. In *Proc. IEEE INFOCOM 2009*.

[18] A. Mtibaa, M. May, M. Ammar, C. Diot. PeopleRank: Combining Social and Contact Information for Opportunistic Forwarding. In *Proc. IEEE Infocom 2010 Mini Conference*, pp. 1-5.

[19] M. Musolesi, C. Mascolo. Designing mobility models based on social network theory. *ACM SIGMOBILE Mobile Computing and Communications Review* 11(3), ACM Press (2007).

[20] L. Pelusi, A. Passarella, M. Conti. Opportunistic Networking: data forwarding in disconnected mobile ad hoc networks. *IEEE Communications Magazine* 44(11), November 2006, pp. 134-141.

[21] J. Su, A. Chin, A. Popivanova, A. Goel, E. de Lara. User mobility for opportunistic ad hoc networking. *Proceedings of IEEE WMCSA*, 2004.

[22] J. Whitbeck, M. Dias de Amorim, V. Conan. Plausible mobility: inferring movement from contacts. *Proceedings of ACM Workshop on Mobile Opportunistic Networking (MobiOpp)*, February 2010.

[23] J. Whitbeck, M. Dias de Amorim, Y. Lopez, J. Leguay, V. Conan. Relieving the Wireless Infrastructure: When Opportunistic Networks Meet Guaranteed Delays. *Proc. IEEE WoWMoM 2011*.

# Vicinity-based DTN Characterization

Tiphaine Phe-Neau*
UPMC Sorbonne Universites
tiphaine.phe-neau@lip6.fr

Marcelo Dias de Amorim*
CNRS and UPMC Sorbonne
Universites
marcelo.amorim@lip6.fr

Vania Conan
Thales Communications
vania.conan@fr.thalesgroup.com

## ABSTRACT

We relax the traditional definition of contact and intercontact times by bringing the notion of *vicinity* into the game. We propose to analyze disruption-tolerant networks (DTN) under the assumption that nodes are in $\kappa$-*contact* when they remain within a few hops from each other and in $\kappa$-*intercontact* otherwise (where $\kappa$ is the maximum number of hops characterizing the vicinity). We make interesting observations when analyzing several real-world and synthetic mobility traces. We detect a number of unexpected behaviors when analyzing $\kappa$-contact distributions; in particular, we observe that in some datasets the average $\kappa$-contact time decreases as we increase $\kappa$. In fact, we observe that many nodes spend a non-negligible amount of time in each other's vicinity without coming into direct contact. We also show that a small $\kappa$ (typically between 3 and 4) is sufficient to capture most communication opportunities.

## Categories and Subject Descriptors

C.2.1 [**Computer-Communication Networks**]: Network Architecture and Design–*Wireless Communication*

## General Terms

Design, Human Factors, Performance

## Keywords

Characterization, Disruption-Tolerant Networks, Opportunistic Networks, Vicinity

## 1. INTRODUCTION

"How much time do two nodes spend within communication range of one another and how long does it take for them to meet again after having left each other?" These questions are central to most works on disruption-tolerant networking (DTN) and concern the problem of evaluating the *contact* and *intercontact* times among nodes [2, 7, 12]. Such analyses have fundamental practical impact, as they serve as a substrate for the design of forwarding strategies that better schedule transmissions based on the history of mobility. Noticeable examples are Prophet [9], Spray and Wait [15], and SSAR [8].

In this paper, we propose a different evaluation of the dynamics of disruption-tolerant networks by integrating the notion of node *vicinity* in the equation. Instead of considering only direct communications among nodes, we suggest extending the notion of "contact" to the zone within a few hops. The impetus for this proposal comes from the observation that a significant fraction of the pairs of nodes remains nearby (within a few hops) when not in direct contact. To this end, we use the $\kappa$-*vicinity*[1] as the basis of our analyses, where $\kappa$ stands for the maximum number of hops separating two nodes. We define and analyze two temporal measures that are the $\kappa$-*contact* and the $\kappa$-*intercontact* times. We analyze their aggregated distributions and keep pairwise analyses for further study.

The interests of relying on an extended view of contacts and intercontacts are manifold. First, one obtains a finer characterization of the network, as we observe in real-world mobility traces that many pairs of nodes are frequently nearby without any direct contact. Second, it becomes easier to tune forwarding protocols as by introducing very little overhead (to discover the neighborhood) a node can discover significant proximity with other nodes. Third, by using short multihop opportunities, end-to-end delays can be decreased.

We analyze the time distributions for the $\kappa$-contact and the $\kappa$-intercontact times using both real-world and synthetic mobility traces and make a number of interesting observations. In a nutshell, we reveal the following findings:

- *Different classes of datasets.* We observe from our analyses that there are basically two different types of behaviors. When extending the contact notion to a node's vicinity, one would expect $\kappa$-contacts duration to increase. However, in some datasets, increasing $\kappa$ leads to a higher probability of having shorter contacts, which is a quite unexpected result. We refer to these patterns respectively as *dense* and *light* distributions.

- *Existing analyses hold for $\kappa$-vicinity.* We confirm that the main observations found in the literature still apply in the context of our study. This means that the main principles of opportunistic communications remain the

*Authors carried out part of the work at LINCS (www.lincs.fr).

[1]Also referred to as the $\kappa$-neighborhood.

same. This is a good point as existing opportunistic protocols can directly benefit from our findings.

- *Close vicinity is enough.* We observe for the datasets we analyze that it is enough to extend the vicinity to a few hops (typically three or four) to capture most of the local communication opportunities. Such a threshold enables low costs for vicinity composition gathering. This also holds for networks with a large diameter.

After introducing and formalizing our proposal, we develop an extended analysis of well-known real-world and synthetic mobility traces. We then discuss the several time distributions and observations we make. Although out of the scope of this paper, we also provide some insights into the applicability of the proposed work in the design of more efficient DTNs.

## 2. BACKGROUND

### 2.1 Related Work

The most intuitive characterization of DTNs relies on the distribution of contact times. A contact occurs when two nodes are within each other's wireless communication range and can perform transmission. We consider networks with bidirectional links. In an early work, Vahdat and Becker investigated the impact of wireless ranges on message delivery [17]. Hui et al. analyzed contacts to derive affinities between individuals and likeliness of meeting [6]. Chaintreau et al. were pioneers in determining the possibilities of efficient transmission in networks through the history of contacts [3]. Contact is not the only meaningful parameter. To make the most of opportunistic communications, understanding intercontact times is also important.

Intercontact distributions indicate when nodes will next be able to transmit data to other devices. In the literature, we observe two main definitions for intercontact. The inter-any-contact notion, meaning the time interval elapsed between two subsequent contacts independently of the identities of the neighbors. The second definition, the pairwise intercontact time, involves a specific pair of nodes. It relates to the time two nodes wait before meeting again after moving away from each other. Leguay et al. thoroughly studied pairwise intercontact distributions in well known experimental datasets. They found these distributions well fitting either log-normal laws or exponential distributions [4]. Chaintreau et al. argued that pairwise intercontact times follow power law distributions over a specific time range [2]. In a similar study, Karagiannis et al. found that pairwise intercontacts fit diptych distributions – power law followed by exponential decay [7]. Recently, Passarella and Conti examined aggregate intercontact times and found them not to be the exact mirror of pairwise intercontact times [12].

### 2.2 Positioning

Although our work is inline with the contributions found in the literature, we propose to characterize DTNs using a different point of view. Previous analyses use one-to-one or one-to-all approaches for characterization. We argue in favor of an in-between approach to leverage a group-to-all vision. In this work, we take into account the immediate vicinity beyond simple contact for every node. We consider a subset of every node's connected component and study the effects of

our group-to-all vision on DTN characterization. Using the $\kappa$-vicinity knowledge, we extend previous analyses to observe the impact of such point of view on DTN understanding.

To study the behavior of contacts within a node's $\kappa$-vicinity, we did not extend a node's wireless range to influence contact possibilities as Vahdat and Becker did. For intercontact patterns, as previously pleaded by Leguay et al., we base our analysis on the pairwise intercontact definition. We rely on our earlier findings that many pairs of nodes, when not in direct contact, remain nearby (within very few hops) [13]. The extension of contact and intercontact to a vicinity notion brings logical variation in previous intercontact and contact analyses. We perform state-of-the-art examination on our extended notions to understand how different our results are when compared to previous pairwise analyses [2, 7].

## 3. FROM THE VIEWPOINT OF VICINITY

Our modern society is tied by the relationships people share with one another. Social network studies showed how people interact based on social ties and how this can be used in networking when they form groups at given times [1, 10]. However, existing DTN protocols still maintain a contact-only vision for their decisions; in other words, they overlook the perspective of nearby nodes. Given people's tendency to colocate in specific places at explicit times, why should we underestimate such information?

### 3.1 The $\kappa$-vicinity

We want to embody the vision a mobile node has of its neighborhood. To this end, we consider the $\kappa$-vicinity of a node instead of only the direct neighbors:

DEFINITION 1. *$\kappa$-vicinity. The $\kappa$-vicinity $\mathcal{V}_\kappa^i$ of node $i$ is the set of all nodes within $\kappa$ hops from $i$.*

We use the term $\kappa$-vicinity to avoid any confusion with the tradition "$n$-hop-neighborhood" terminology. We assume that the $n$-hop-neighborhood indicates the nodes *exactly* at $n$-hops, while $\kappa$-vicinity gathers all the nodes *up to* $\kappa$ hops. In Figure 1, we illustrate the 2-vicinity of node $i$.

The $\kappa$-vicinity brings the immediate surrounding knowledge. This is an interesting point of view for opportunistic networks because it extends a node's knowledge to immediately useable communication opportunities. The $\kappa$-vicinity empowers a node's reach in the network [14].

### 3.2 $\kappa$-contact and $\kappa$-intercontact

We are now ready to make the necessary definitions for the rest of our work.

DEFINITION 2. *$\kappa$-contact. Two nodes are in $\kappa$-contact when they dwell within each other's $\kappa$-vicinity, with $\kappa \in \mathbb{N}^*$. More formally, two nodes $i$ and $j$ are in $\kappa$-contact when $\{i \in \mathcal{V}_\kappa^j\} = \{j \in \mathcal{V}_\kappa^i\}$. In other words, a contemporaneous path of length at most $\kappa$ links $i$ and $j$.*

We also need to grasp the intercontact observations for our vicinity viewpoint. The literature definition of mere intercontact is when two nodes are not in contact. Therefore, we consider $\kappa$-intercontact when two nodes are not in $\kappa$-contact. These are complementary notions. Another way to see it is as follows: if node $i$ maintains knowledge about its $\kappa$-vicinity, it is in $\kappa$-intercontact with any node beyond its

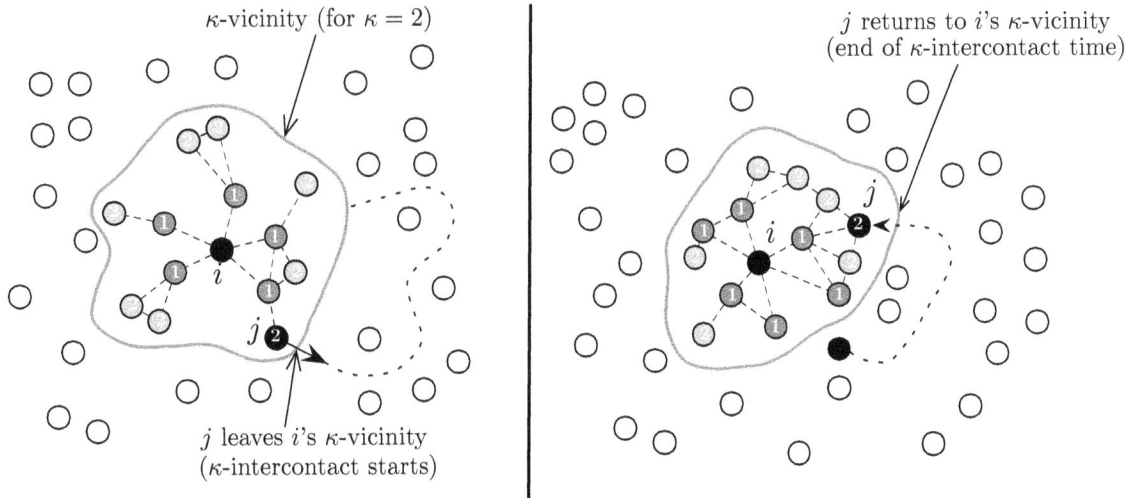

Figure 1: Node $i$'s $\kappa$-vicinity and the $\kappa$-intercontact phenomenon. For the sake of clarity, we only display connectivity links and shortest hop distance from $i$ within the $\kappa$-vicinity.

$\kappa$-vicinity. In Figure 1, node $j$ leaves $i$'s $\kappa$-vicinity and then gets back some time later, characterizing a $\kappa$-intercontact interval.

DEFINITION 3. $\kappa$-**intercontact**. *Two nodes are in $\kappa$-intercontact while they do not belong to each other's $\kappa$-vicinity. Formally speaking, two nodes $i$ and $j$ are in $\kappa$-intercontact when $\{i \notin \mathcal{V}_\kappa^j\}$ or $\{j \notin \mathcal{V}_\kappa^i\}$. There is no path of length $\kappa$ or less linking $i$ and $j$.*

Note that 1-contact matches the contact notion and 1-intercontact corresponds to usual binary intercontact.

## 4. DATASETS

To perform our analysis, we selected real-life datasets and synthetic models displaying specific scenarios. Each of them embeds characteristics of real-life patterns that DTN wants to leverage. Real-life measurements used devices capturing other devices presence within a 10m wireless range.

**Infocom05** measurement was held during a 5 day conference in 2005 [2]. 41 attendees carried iMotes collecting information about other iMotes nearby. We study a 12-hour interval bearing the highest networking activity. Each iMote probes its environment every 120 seconds. *Infocom05* represents a professional meeting framework.

**Rollernet** had 62 participants measuring their mutual connectivity with iMotes while they where riding a dominical rollerblading tour during 3 hours in Paris [16]. Researchers set devices to send beacons every 30 seconds. These measurements show a specific sport gathering scenario.

**Unimi** is a dataset captured by students, faculty members, and staff from the University of Milano in 2008 [5]. They involved 48 persons with special devices probing their neighborhood every second. *Unimi* provides a scholar and working environment scenario.

**RT** is a mobility model correcting flaws from the Random Waypoint model [11]. We sampled the behavior of 20 nodes following this model on a surface of 50 x 60 m$^2$ with speed

between 0 and 7 m/s and a 10m wireless range for vicinity sensing.

**Community** is a social-based mobility model [10]. It tends to colocate socially-related nodes in specific locations at the same time like groups of friends would do. We simulated 50 nodes with a 10m wireless range on a 1,500 x 2,500 m$^2$ plane during 9 hours.

## 5. INTERCONTACT DISTRIBUTIONS

Intercontact patterns in DTN sprang many inspiring analyses as seen in Section 3. Studying intercontact duration distributions helps researchers quantify how long a node will have to wait before its next encounter. Figure 2 represents aggregated complementary cumulative density function (CCDF) of binary (traditional) intercontact and respective $\kappa$-intercontact durations for every pair of nodes. These CCDFs indicate the probability of a $\kappa$-intercontact lasting longer than $t$ seconds.

### 5.1 Binary intercontact

As Karagiannis et al. observed, we also find that all binary intercontact CCDFs follow a straight line up to a knee point when both $x$-axis and $y$-axis are on logarithmic scale. This implies power laws for each binary intercontact distributions until the observed knee point also known as the characteristic time. In *Unimi*, we observe a knee point for binary intercontact at around 50,000 seconds. After plotting distributions with a linear scale on the $x$-axis and maintaining the log scale on the $y$-axis, we also observe that distributions can be approximated by a straight line beyond the knee point. In Figure 2, the phenomenon is clear for *Unimi*. This hints exponential decays for distribution tails. Observations on binary intercontact match results of previous studies.

### 5.2 $\kappa$-intercontact

As of $\kappa$-intercontact distributions, we find their general overlook to be quite similar to their respective binary intercontact distributions except for *Community*. $\kappa$-intercontacts display partial displacement after some point with a sharper

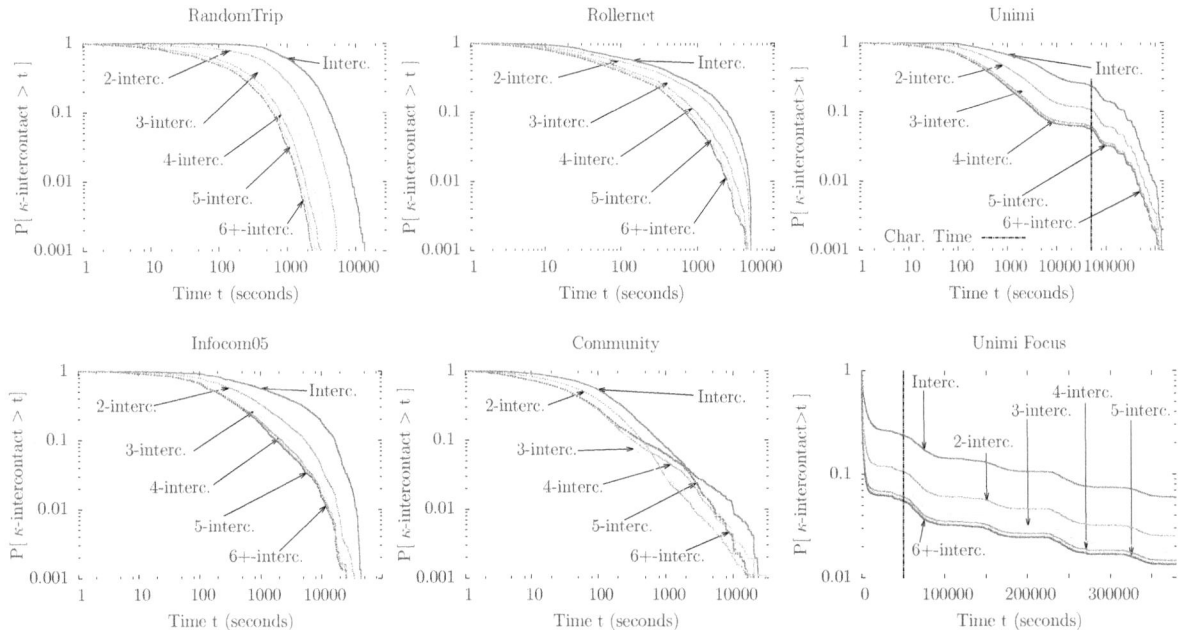

Figure 2: $\kappa$-intercontact distributions. Apart from Community, nodes display a lower probability of obtaining $\kappa$-intercontact intervals lasting longer than t seconds for high $\kappa$. On average, $\kappa$-vicinity reduces $\kappa$-intercontact times. Distributions follow power laws up to a characteristic time and display exponential decay afterwards. All $\kappa$-intercontact distributions knee point concord. Community has inconsistent $\kappa$-vicinity patterns for $\kappa \geq 3$. Interc. stands for Binary intercontact (logscale on both axes except Unimi Focus which is linear-log).

Table 1: $\kappa$-intercontact average duration in seconds.

| Dataset | $\kappa$ | | | | | |
|---|---|---|---|---|---|---|
| | 1 | 2 | 3 | 4 | 5 | 6+ |
| RT | 1,874.3 | 772.2 | 415.7 | 291.5 | 238.4 | 213.1 |
| Rollernet | 738.3 | 555.3 | 412.2 | 328.2 | 273.6 | 242.5 |
| Unimi | 66,434.8 | 28,687.6 | 19,529.0 | 16,585.3 | 15,534.9 | 15,110.3 |
| Infocom05 | 4,930.9 | 1,752.0 | 1,111.5 | 916.8 | 850.3 | 823.5 |
| Community | 525.3 | 232.4 | 193.5 | 262.2 | 317.1 | 295.9 |

slope for each curve. The larger the $\kappa$ parameter, the more important the bottom left shift for each distribution. The concept of $\kappa$-neighborhood logically reduces $\kappa$-intercontact times. The wider a node's vicinity knowledge, the later this latter detects a node's departure from its vicinity and the quicker it detects its comeback resulting in shorter $\kappa$-intercontact durations. We see that for $\kappa \geq 6$, CCDFs aggregate.

An interesting remark is how $\kappa$-intercontact distributions exhibit the same properties as binary intercontacts. They follow power laws until a specific point (the characteristic time) and then carry exponential decay. In Figure 2, beyond 50,000 seconds, *Unimi*'s 2-intercontact curve is a vertical shift of the binary intercontact CCDF. The same occurs for further $\kappa$-intercontacts. However, an important information is that the knee point found for binary intercontact corresponds to changing points for $\kappa$-intercontact distributions. In *Unimi*, $\kappa$-intercontact curves ($\kappa \geq 2$) quickly decrease after the characteristic time found at 50,000 seconds.

Table 1 displays average intercontact duration and Table 2 the number of intercontacts intervals for each dataset. Except for *Community*, the higher $\kappa$ gets, the lower the aver-

age $\kappa$-intercontact length. This enforces our rational expectations of $\kappa$-vicinity reducing $\kappa$-intercontact duration with higher $\kappa$. We remarked decreasing cumulated $\kappa$-intercontact times for each $\kappa$. We also observe a logarithmic growth in the number of $\kappa$-intercontact intervals.

## 5.3 Observations

The *Community* dataset stands out because of its non-monotonic average $\kappa$-intercontact duration and evolution of the number of intervals. When the average length grows, the number of $\kappa$-intercontact intervals decreases. This still results in a decreasing cumulated $\kappa$-intercontact duration for each $\kappa$. It enforces our first thoughts in the benefits of $\kappa$-vicinity for $\kappa$-intercontact times.

Under the assumption that nodes in the vicinity dwell within low delay reach, $\kappa$-intercontact duration decreases with larger $\kappa$, strengthening our belief that $\kappa$-neighborhood is beneficial to DTN protocols. The fact that characteristic time in all intercontact distributions corresponds is also an important finding. It could help protocols like Prophet, Spray-and-Wait, or SSAR maintain their actual intercontact-

Table 2: $\kappa$-intercontact number of intervals.

| Dataset | $\kappa$ | | | | | |
|---|---|---|---|---|---|---|
| | 1 | 2 | 3 | 4 | 5 | 6+ |
| RT | 2,264 | 5,041 | 8,258 | 10,516 | 11,862 | 12,629 |
| Rollernet | 2,529 | 7,460 | 12,357 | 15,789 | 17,622 | 18,422 |
| Unimi | 21,737 | 57,085 | 86,009 | 102,495 | 109,406 | 112,323 |
| Infocom05 | 3,727 | 11,028 | 15,338 | 16,774 | 17,186 | 17,117 |
| Community | 3613 | 11,561 | 8,034 | 4,505 | 3,660 | 3,477 |

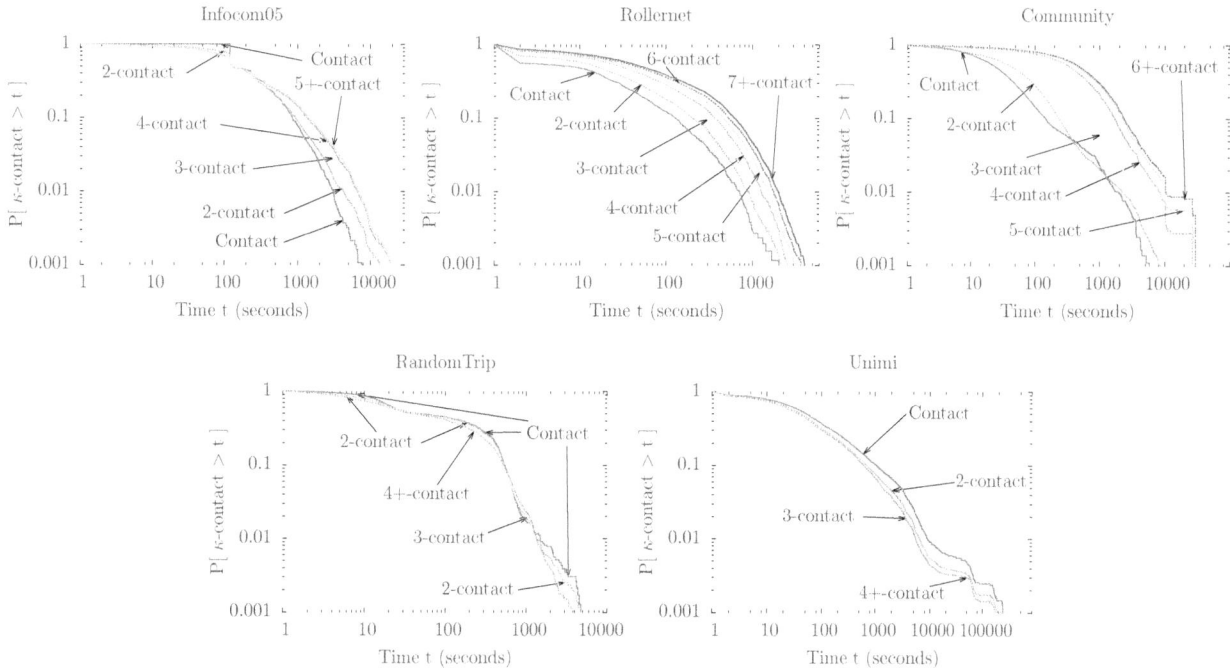

Figure 3: $\kappa$-contact distributions. There are two major patterns: (i) *dense distributions* where CCDFs having larger $\kappa$ suffer a top right shift and a smoother slope than smaller $\kappa$, and (ii) *light distributions*, where all $\kappa$-contact distributions for $\kappa \geq 4$ aggregate and present a slight bottom left shift compared to the contact distribution (logscale on both axes).

based approach and extend them to their vicinity to benefit from shorter $\kappa$-intercontact times.

# 6. CONTACT AND $\kappa$-CONTACT ANALYSES

Contact is the main feature for opportunistic mobile networks. Analyzing its distribution gives us insights into how protocols can benefit from these contact opportunities, as $\kappa$-contacts happen to be an extension of the strict contact definition. Instead of considering contact between neighbors at a 1-hop distance only, we analyzed the potential of transmission to nearby nodes within the $\kappa$-vicinity. These paths enable low delay transmissions and a better neighborhood reach for a network node.

## 6.1 $\kappa$-contact duration distributions

In Figure 3, we display aggregated CCDFs of contact alongside $\kappa$-contact duration for every pair of nodes in each experiment. These CCDFs indicate the probability of a $\kappa$-contact lasting longer than $t$ seconds.

For *Infocom05* and *Rollernet*, their CCDFs maintain comparable aspects. We observe a small upper right shift for larger values of $\kappa$. As the $\kappa$-contact notion increases the

node's vicinity scope, any nearby node may stay within the considered node's coverage longer than with a shorter sight vision. The higher the $\kappa$, the higher the probability of having longer $\kappa$-contact intervals duration. Above scanning granularity, lower $\kappa$ results in curves with a sharper slope than curves of longer $\kappa$-contacts.

Like *Infocom05* and *Rollernet*, for $\kappa \geq 3$, *Community*'s $\kappa$-contact CCDFs bear the same overall outlook with a sharper slope for smaller $\kappa$. For 1- and 2-contact CCDFs, we hint an interesting phenomenon. We find two junctions around 400 seconds and another at 1,050 seconds. Opposed to our previous expectations, we have a better chance of getting contact of duration $D \in [400; 1, 050]$ seconds than 2-contact slots of the same duration.

For *RT* and *Unimi*, their 1-contacts bear different behaviors than $\kappa$-contacts when $\kappa \geq 2$. As hinted in the *Community* dataset, for some times $\kappa$, the probability of obtaining contacts slots lasting longer than $t$ seconds is higher than the probability for the same $t$ in other datasets. In Figure 3, this phenomenon clearly appears for *Unimi*. In *RT*, $t = [0; 500] \cup [1, 050; 10, 000]$ seconds. In *Unimi*, the assertion is valid for the whole distribution. For $\kappa \geq 3$, CCDFs

Table 3: $\kappa$-contact average duration in seconds.

| Dataset | $\kappa$ | | | | | | |
|---|---|---|---|---|---|---|---|
| | 1 | 2 | 3 | 4 | 5 | 6 | 7+ |
| Infocom05 | 371.1 | 406.4 | 492.8 | 561.2 | 597.5 | 630.5 | 653.0 |
| Rollernet | 47.2 | 73.5 | 97.7 | 125.7 | 156.0 | 184.3 | 211.6 |
| Community | 96.3 | 138.7 | 358.6 | 751.9 | 1,000.9 | 1,123.21 | 1,135.5 |
| RT | 201.8 | 200.2 | 184.7 | 182.2 | 182.1 | 182.1 | 182.4 |
| Unimi | 1.324.6 | 901.2 | 820.7 | 796.7 | 791.5 | 801.5 | 798.2 |

Table 4: $\kappa$-contact number of intervals.

| Dataset | $\kappa$ | | | | | | |
|---|---|---|---|---|---|---|---|
| | 1 | 2 | 3 | 4 | 5 | 6 | 7+ |
| Infocom05 | 3,735 | 11,071 | 15,412 | 16,870 | 17,288 | 17,221 | 17,117 |
| Rollernet | 5,106 | 15,410 | 25,200 | 32,630 | 36,286 | 38,110 | 38,586 |
| Community | 3,629 | 11,612 | 8,127 | 4,627 | 3,798 | 3,627 | 3,598 |
| RT | 2,316 | 5,146 | 8,385 | 10,645 | 11,992 | 12,759 | 13,128 |
| Unimi | 10,875 | 28,550 | 43,019 | 51,271 | 54,733 | 56,193 | 56,782 |

aggregate into an unique one. 2-contact distribution is a mixed behavior between 1-contact and larger values of $\kappa$.

## 6.2 Density related behavior

Due to the social nature of *Community*'s functioning, specific nodes tend to remain together and bring a high density around popular nodes. *Rollernet* is a dense sport setting and *Infocom05* has selective meeting points in the conference. They all exhibit an important node-centered density, whereas *Unimi* and *RT* bear light density around each nodes. The local density parameter may explain the difference between the $\kappa$-contact behaviors.

Figure 4 illustrates a situation detailing the unexpected behavior of $\kappa$-contact distributions in light settings. Lighter densities limits geographical $\kappa$-vicinity coverage and induces smaller $\kappa$-contact intervals. Dense settings ignite distributions like *Infocom05*, *Rollernet*, and *Community* and will be henceforth mentioned as *dense distributions*. Low density settings like *RT* and *Unimi* enable the second type of distributions mentioned as *light distributions*.

## 6.3 Average duration and number of intervals

In Table 3, we displayed the average duration of $\kappa$-contact intervals and in Table 4 the number of $\kappa$-contact slots for each of our five experiments. Two main behaviors arise. On the one hand, for *Infocom05*, *Rollernet*, and *Community*, we find an impressive continuous growth of average slots duration for every $\kappa$. On the other hand, *RT* and *Unimi* show the opposite evolution concerning average $\kappa$-contact duration. An increase in $\kappa$ brings increased average $\kappa$-contact lengths.

For most datasets, we also find a logarithmic growth of the number of $\kappa$-contact intervals. Consequently, the number of intervals balances their length shortening. This testifies the growth in cumulated $\kappa$-contact durations in all datasets. Despite results observed in the previous section for *RT* and *Unimi*, for all our datasets, we find that larger values of $\kappa$ increase the overall $\kappa$-contact duration and modify its distribution. The main difference lies in the fact that *Infocom05*, *Rollernet*, and *Community* experience longer $\kappa$-contacts for large $\kappa$ than *RT* and *Unimi*, which have more shorter $\kappa$-

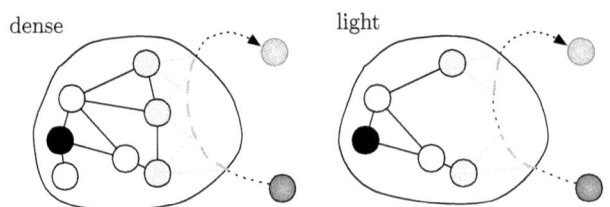

Figure 4: Density related behavior for $\kappa$-contact. Density modifies the coverage zone of a node's $\kappa$-vicinity. For dense settings, we have a long continuous $\kappa$-contact interval. For light situations, we obtain two smaller $\kappa$-contacts for the same walk.

contacts. In any case, both types have longer cumulated $\kappa$-contact times and it grows with $\kappa$.

## 6.4 Observations

We have seen how $\kappa$-contact distributions predominantly exhibit two behaviors: light or dense distributions. Dense distributions follow our logical expectations. These distributions have sharper slope for lower $\kappa$ and therefore a stronger demarcation among them than the next variety. Light distributions show $\kappa$-contact distributions with comparable behaviors and no major demarcations. They quickly aggregate into a unique distribution above $\kappa \geq 4$. For these distributions, contrary to our primary beliefs, the probability of getting $\kappa$-contacts longer than $t$ seconds is higher for shorter values of $\kappa$ and contact durations.

However, for all measurements, the number of $\kappa$-contact intervals increases with every $\kappa$ and springs a longer cumulated $\kappa$-contact time. Dense distribution obtains more large $\kappa$-contact intervals whereas light distribution has more short $\kappa$-contact intervals. Knowing which distribution fits, either light or dense, to a given situation modifies the way a protocol should consider its $\kappa$-vicinity. Adapting a routing technique to dense or light $\kappa$-contact distributions accordingly may help nodes leverage their $\kappa$-vicinity more efficiently than what is currently done.

Table 5: Average number of neighbors $\mathcal{D}_\kappa^i$ in a node's $\kappa$-vicinity (whole dataset duration).

| Dataset | $\kappa$ | | | | | | | |
|---|---|---|---|---|---|---|---|---|
| | 1 | 2 | **3** | **4** | 5 | 6 | 7 | 8+ |
| Community | 2.0 | 4.1 | 4.6 | 4.9 | 4.9 | 4.9 | 4.9 | 4.9 |
| RT | 2.0 | 3.2 | 4.7 | 6.7 | 7.4 | 7.8 | 7.9 | 8.0 |
| Infocom05 | 2.3 | 3.8 | 5.5 | 6.0 | 6.3 | 6.4 | 6.4 | 6.4 |
| Rollernet | 1.8 | 3.2 | 4.7 | 5.7 | 6.3 | 6.7 | 6.9 | 7.1 |
| Unimi | 0.15 | 0.25 | 0.31 | 0.35 | 0.37 | 0.38 | 0.38 | 0.38 |

## 7. $\kappa$-VICINITY ANALYSIS

Where most studies consider only the possibilities of contacts, using a node's vicinity sounds appealing to reduce $\kappa$-intercontacts and increase $\kappa$-contact times. With the $\kappa$-vicinity, we can measure the potential of such nearby companions in terms of opportunistic communications. Yet, we can wonder up to which point a node should survey its vicinity to leverage it.

### 7.1 Density

To mirror a node's specific $\kappa$-vicinity density, for each node, let $\mathcal{D}_\kappa^i$ be the density of nodes around $i$, obtained as

$$\mathcal{D}_\kappa^i = \frac{card(\mathcal{V}_\kappa^i)}{\tau}, \qquad (1)$$

where $card(\mathcal{V}_\kappa^i)$ is the number of nodes in $i$'s $\kappa$-vicinity and $\tau$ is the experiment duration. $\kappa$-density internal composition influences a node's $\kappa$-vicinity behavior. The more $\kappa$-contacts a node has, the more chances it has of getting $\{\kappa + 1\}$-contacts. In Table 5, we present the average $\mathcal{D}_\kappa^i$. For datasets with participants moving slow and steady like *Infocom05*, *Unimi*, and *Community*, above a certain limit $\mathcal{D}_\kappa^i$ does not increase anymore and is limited by the network diameter. More dynamic or inconsistent patterns – *RT* and *Rollernet* – display logarithmic increase in $\mathcal{D}_\kappa^i$. For all cases, we verify $\mathcal{V}_\kappa$ growth with $\kappa$ indicating the presence of nearby nodes useable as relays for $\kappa$-contact.

For any datasets, observing contacts only shows limited $\mathcal{D}_\kappa^i$. While observing the $\kappa$-vicinity up to a few hops – $\kappa = \{3, 4\}$ – increases $\mathcal{D}_\kappa^i$ by at least doubling it or even tripling it. For $\kappa > 4$, the increase rate is less striking or even null. Nevertheless, longer $\kappa$-contacts in terms of path length may not be interesting because of potential path inconsistency due to all relays movements. Monitoring $\kappa$-vicinity up to a $\{3, 4\}$ threshold brings most of the local density a node can use.

### 7.2 Neighbors beyond contacts

An interesting situation occurs when pairs of nodes do not come into contact but belong to each other's $\kappa$-vicinity. Usual protocols miss this knowledge by overlooking the potential of nearby nodes. To analyze the impact of such situations, we studied the closest distance between nodes for all pair of nodes.

For *Unimi* and *Infocom05*, we find that respectively 92% and 91% of pair of nodes do come in contact. This can be explained by the datasets nature where people are coworkers and have to meet to exchange ideas. However, we find that even there, some nodes do not find themselves closer than a 2-hop distance respectively for 6% and 7% of them. Other

datasets deprived of the specific aim of meeting each other like *Rollernet* and *Community* show that contact only represent 31% and 42% of the lowest distances. There, respectively more than 51% and 46% stay at the closest between 2 and 4-hop distances. In *RT*, all pair of nodes come into contact at one point or another. By observing the $\{3, 4\}$-vicinity, we manage to monitor additional situations of non-contact between nodes. As a result, we catch most pairwise $\kappa$-contacts occurring in a node's vicinity with only a threshold $\kappa = \{3, 4\}$.

## 8. IMPLICATIONS

**Opportunistic Protocols.** DTN protocols chose to leverage the obvious contacts – binary intercontact patterns. While they may appear sufficient to elaborate routing schemes, they ignore nearby communications possibilities. As DTN rely on human mobility patterns to generate encounters and topological proximity, we should make use of hot places in a map and hubs on the connectivity plane. Gathering a node's surrounding situation via the $\kappa$-vicinity knowledge can help us do so. We show that observing a node's $\kappa$-vicinity improves both contact opportunities and intercontact durations. Moreover, in Section 7, we explain how observing a node's $\kappa$-vicinity with $\kappa = \{3, 4\}$ is enough to be aware of most pairwise activity in the vicinity beyond contacts and to benefit from local densities.

**Mobility Models.** Musolesi et al. based their mobility model on social network theory [10]. Their model takes into account colocating patterns by mean of social attractiveness. Their intent, with HCMM proposed by Boldrini et al [1], is one of the most sensible we have seen in terms of binding synthetical models and social patterns. Still, they limit their approach to contact patterns which results in incoherent $\kappa$-contact and $\kappa$-intercontact distributions. In Figure 3, contact and 2-contact distributions intertwine whereas the $\kappa$-contact ($\kappa \geq 3$) do not and present the same demarcation as other *dense distributions*. In Figure 2, binary intercontact and 2-intercontact CCDFs present expected behaviors. We find 3+-intercontact behavior inconsistent. We warn users when using such mobility models, while issued traces respect essential social patterns, they may mislead users on other incidental parameters like $\kappa$-vicinity behaviors.

## 9. CONCLUSION

We propose a DTN characterization based on the vicinity viewpoint. Our motivation comes from the fact that most DTN protocols ignore their vicinity beyond one hop. We confirm previous results of Karagiannis et al. with regard to aggregated $\kappa$-intercontact behaviors, meaning that they follow power laws up to a given time and experience exponential decay afterwards. This allows current DTN protocols to leverage their $\kappa$-vicinity without too much change in their functioning. We also found that $\kappa$-contact distributions globally follow two patterns which are density related. Dense environments provide logical results of $\kappa$-contact duration extension with higher $\kappa$ whereas light settings display inverted paradoxical patterns. Protocols should be aware of these patterns and treat them accordingly to benefit from this knowledge. Finally, we showed how limiting a node's awareness to their $\{3,4\}$-vicinity is enough to benefit from most $\kappa$-vicinity advantages. As a next step, we plan on analyzing pairwise $\kappa$-vicinity behaviors on a strict pairwise level to enable better identification of peculiar events between nodes. We also would like to dive deeper into the $\kappa$-contact notion and understand the different path types resulting in $\kappa$-contact.

## Acknowledgments

This work is partially supported by the ANR project CROWD under contract ANR-08-VERS-006.

## 10. REFERENCES

[1] C. Boldrini and A. Passarella. HCMM: Modelling spatial and temporal properties of human mobility driven by users' social relationships. *Computer Communications*, 33(9):1056 – 1074, 2010.

[2] A. Chaintreau, P. Hui, J. Crowcroft, C. Diot, R. Gass, and J. Scott. Impact of Human Mobility on Opportunistic Forwarding Algorithms. *IEEE Transactions on Mobile Computing*, 6(6):606–620, 2007.

[3] A. Chaintreau, A. Mtibaa, L. Massoulie, and C. Diot. The Diameter of Opportunistic Mobile Networks. In *ACM CONEXT*, 2007.

[4] V. Conan, J. Leguay, and T. Friedman. Characterizing Pairwise Inter-contact Patterns in Delay Tolerant Networks. In *Autonomics*, 2007.

[5] S. Gaito, E. Pagani, and G. P. Rossi. Fine-Grained Tracking of Human Mobility in Dense Scenarios. In *IEEE SECON poster session*, 2009.

[6] P. Hui, J. Crowcroft, and E. Yoneki. BUBBLE Rap: Social-Based Forwarding in Delay-Tolerant Networks. *IEEE Transactions on Mobile Computing*, 10:1576–1589, 2011.

[7] T. Karagiannis, J.-Y. Le Boudec, and M. Vojnović. Power Law and Exponential Decay of Inter Contact Times between Mobile Devices. In *ACM MOBICOM*, 2007.

[8] Q. Li, S. Zhu, and G. Cao. Routing in Socially Selfish Delay Tolerant Networks. In *IEEE INFOCOM*, 2010.

[9] A. Lindgren, A. Doria, E. Davies, and S. Grasic. Probabilistic Routing Protocol for Intermittently Connected Networks draft-irtf-dtnrg-prophet-09, 2011. DTN Research Group Internet-Draft.

[10] M. Musolesi and C. Mascolo. Designing Mobility Models based on Social Network Theory. *SIGMOBILE Mob. Comput. Commun. Rev.*, 11:59–70, July 2007.

[11] S. Pal Chaudhuri, J.-Y. Le Boudec, and M. Vojnovic. Perfect Simulations for Random Trip Mobility Models. In *IEEE INFOCOM*, 2005.

[12] A. Passarella and M. Conti. Characterising aggregate inter-contact times in heterogeneous opportunistic networks. In *IFIP NETWORKING*, 2011.

[13] T. Phe-Neau, M. Dias de Amorim, and V. Conan. Fine-Grained Intercontact Characterization in Disruption-Tolerant Networks. In *IEEE ISCC*, 2011.

[14] T. Phe-Neau, M. Dias de Amorim, and V. Conan. Using Neighborhood Beyond One Hop in Disruption-Tolerant Networks, 2011. http://arxiv.org/abs/1111.0882v1.

[15] T. Spyropoulos, K. Psounis, and C. S. Raghavendra. Spray and Wait: An Efficient Routing Scheme for Intermittently Connected Mobile Networks. In *ACM SIGCOMM workshop on Delay-tolerant networking*, 2005.

[16] P.-U. Tournoux, J. Leguay, F. Benbadis, J. Whitbeck, V. Conan, and M. D. de Amorim. Density-Aware Routing in Highly Dynamic DTNs: The RollerNet Case. *IEEE Transactions on Mobile Computing*, 10:1755–1768, 2011.

[17] A. Vahdat and D. Becker. Epidemic Routing for Partially Connected Ad Hoc Networks. Technical report, Duke University, 2000.

# Performance of Collaborative GPS Localization in Pedestrian Ad Hoc Networks

Vladimir Vukadinovic
Disney Research Zurich
8092 Zurich, Switzerland
vvuk@disneyresearch.com

Stefan Mangold
Disney Research Zurich
8092 Zurich, Switzerland
stefan@disneyresearch.com

## ABSTRACT

In ad hoc networks without static nodes that could be used as reference points, mobile handhelds must rely on their GPS receivers to enable location-aware services. By sharing their position estimates using short range radios, neighboring devices may suppress unnecessary GPS activations in order to reduce energy consumption. We describe and evaluate two collaborative GPS localization protocols based on substitution and averaging of position estimates. The evaluation focuses on entertainment park scenarios and relies on realistic simulations to capture the mobility of park visitors. We demonstrate that the simple collaboration protocols, which do not require distance estimation between the neighbors, may provide significant energy savings. We discuss the impact of device density and provide guidelines for choosing the transmission range of their radio interfaces.

## Categories and Subject Descriptors

C.2.1 [**Computer–Communication Networks**]: Network Architecture and Design–*Wireless communication.*

## General Terms

Algorithms, Performance, Experimentation.

## Keywords

GPS localization; energy consumption; ad hoc networks; mobility.

## 1. INTRODUCTION

Unlike in infrastructure-based wireless networks, communication in ad hoc networks relies on dynamically-created multi-hop routes composed of direct links between neighboring nodes, which forward data on behalf of each other. This communication mode is useful when network infrastructure (cellular base stations, WLAN access points) is not available or cannot be used due to the cost or logistic reasons. It can also be used to supplement sparsely deployed infrastructure networks with partial coverage. Examples of ad hoc network architectures include mobile ad hoc networks (MANETs) and delay/disruption tolerant networks (DTNs). Today, the use of the ad hoc mode on handheld devices, such as smartphones, is hindered by the high energy consumption of Wi-Fi radios, the short range and slow pairing procedure of Bluetooth radios, the complexity of ad hoc network setup, and lack of support in popular operating systems (e.g. iOS and Android).

However, as Wi-Fi Direct [1] and Bluetooth 4.0 (aka Bluetooth Smart) [2] are starting to penetrate the market, it is likely that such obstacles will diminish. Nokia's Instant Community [3], Apple's iGroups [4] currently rely on proprietary solutions to enable ad hoc communication between smartphones.

Depending on the density and mobility of nodes, an ad hoc network may exhibit varying degrees of partitioning and, therefore, cannot provide stringent quality of service guarantees. However, if carefully designed, a variety of application can be provided in such networks. For example, ad hoc networks can support mobile multi-player games, mobile advertising, multimedia sharing, and participatory sensing. Some of these applications are location-based and require knowledge of user's location with different levels of accuracy. In infrastructure-based networks, the location can be determined using multilateration algorithms, using infrastructure nodes at known locations as reference points. In ad hoc networks, static infrastructure nodes are not available or they are too sparse to be used for multilateration. Without WLAN access points, GSM/UMTS cell towers can only provide very coarse localization, with errors as large as 300 m. GPS localization becomes often the only way to localize a handheld device. However, GPS is power hungry: A continuously active GPS receiver alone may drain a battery on a smartphone in a few hours. Our results show that its consumption is significant even if sampled every few minutes.

In this paper, we consider collaboration between neighboring devices to reduce the energy consumption of GPS receivers. A device may combine position information received from the neighbors with its own to obtain a new position estimate without relying on GPS. Only if a position estimate with acceptable error/confidence cannot be obtained from the neighbors, a device will trigger its GPS receiver. We refer to this type of localization as *collaborative GPS localization*. The collaboration may not only reduce the GPS energy consumption, but it may also increase the positioning accuracy: Even static and collocated devices often obtain different position information from GPS due to multipath effects and/or because their receivers have different sensitivities and lock-on different sets of satellites. If the position information is shared between the devices, each device may refine its original position estimate. However, it is not straightforward to conclude that collaborative GPS localization is always beneficial. While it may reduce the consumption of GPS receivers, the collaboration increases the consumption of wireless interfaces. Also, if a GPS receiver is not sampled regularly, it may require a new satellite lock-on procedure, which is a lengthy and power consuming procedure. Furthermore, the overall energy consumption and accuracy of collaborative GPS localization depends strongly on the density and mobility of devices in an area and on the transmission range of their radios.

Our target scenarios assume large crowds of people/tourists visiting areas such as entertainment theme parks, zoos, open-air archeological parks, or nature reserves. A number of location-based services offered to the visitors rely on wireless communication [5]. However, it cannot be assumed that such areas are fully covered with a wireless (e.g. Wi-Fi) infrastructure. Rolling out extensive infrastructure in a theme park, for example, is not an easy task: The largest parks are comparable in size with big cities (Walt Disney World Resort in Florida spans over ~100 km$^2$). Problems go beyond the obvious deployment and maintenance costs. For example, access points and antennas may be too visible to guests and, therefore, interfere with artistic intentions. Therefore, services must rely on spotty Wi-Fi coverage (if any), ad hoc communication between visitors' devices, and GPS localization. We describe a protocol for collaborative GPS localization that decides when a device should provide/request location information to/from its neighbors. It then calculates a new position estimate based on the input from the neighbors. We explore if such protocol can be used to reduce the energy consumption of GPS receivers. Our evaluation is based on detailed simulations, where the mobility of people/devices is driven by mobility traces collected in a theme park [5].

The reminder of this paper is organized as follows: Section 2 provides an overview of related work. Energy consumption models for GPS receivers and wireless interfaces are described in Section 3. Section 4 describes methods to measure and track position errors on smartphones and introduces an error model used in our simulations. Section 5 presents our collaborative localization protocols. The simulation setup and performance results are presented in Sections 6 and 7, respectively. Section 8 concludes the paper.

## 2. RELATED WORK

Collaborative (cooperative) localization has been first proposed for robot and sensor networks. Only recently it has been considered for other types of wireless networks. A general overview of cooperative localization techniques is provided in [6]. In robotics literature, collaborative localization refers to the problem of fusing relative position measurements between mobile robots with their odometry measurements. Most of the proposed solutions require centralized processing [7][8], although distributed algorithms have also been proposed [9].

In sensor networks, cooperating devices are typically densely deployed and static. A subset of nodes may be equipped with GPS receivers and/or able to obtain their absolute locations by other means. These devices serve as anchors for other devices, which rely on ranging and multilateration to determine their locations. Most of the solutions proposed for static sensor networks are not applicable to mobile networks due to frequent changes in topology. Therefore, [10] proposes a collaborative localization scheme by which a mobile sensor node may refine its location estimate through sporadic encounters with other nodes. In [11], the authors evaluate the benefits of using radio ranging and location sharing among GPS-enabled sensors mounted on cow collars for cattle tracking and virtual fencing applications. Two collaborative localization algorithms that rely on accurate ranging between neighbors using ZigBee and UWB radios have been described in [12]. Similarly, [13] considers cooperative WLAN-based indoor positioning for groups of people moving in clusters. ZigBee radios are used for proximity detection and communication within the clusters.

In sensor networks, energy-efficient communication between cooperating nodes can be achieved by means of ultra-low power radios. However, if communication relies on radios available on commodity phones, such as Wi-Fi and Bluetooth radios, the collaboration may incur significant energy overhead. Cooperative

positioning for Wi-Fi devices is considered in [14]. Received signal strength (RSS) measurements towards access points, which are used as reference points, are supplemented with RSS measurements towards neighboring mobile terminals and used as inputs to an algorithm that calculates the position. A similar scheme is proposed and evaluated in [15]. In [16], the authors consider collaboration among mobile nodes capable of localizing themselves using either GPS or pedestrian dead reckoning. The focus in [14]-[16] is on localization accuracy; energy consumption is not addressed. Simulations in [15], [16] employ simple random walk mobility models. Our work is closely related [17], which considers Bluetooth communication to synchronize GPS positions of neighboring devices to reduce the number of GPS activations. To evaluate the benefits of the proposed Bluetooth-based Position Synchronization (BPS) protocol, two phones were placed in a bag and carried around for two days. Therefore, both phones had the same GPS signal availability and were constantly within each other's range. Depending on the mobility and density of devices, this simplistic evaluation could underestimate or overestimate the potential energy savings of the protocol in real-world scenarios.

Several works have proposed to use accelerometers, compasses, microphones, and other low-power sensors on phones [17]-[21] as well as the knowledge of habitual mobility [22] to adapt the duty cycles and suppress unnecessary activations of GPS receivers. Such approaches are complementary to the collaborative GPS localization and can be used to further reduce the energy consumption. However, they can be very unreliable: For example, theme park visitors engage in many activities that activate the accelerometers (e.g. ride a roller-coaster), but do not result in mobility that needs to be tracked.

## 3. ENERGY CONSUMPTION

In this section, we describe models of the energy consumption of GPS receivers and wireless interfaces, which we later use to evaluate the performance of collaborative GPS localization.

### 3.1 Energy Consumption of GPS Receivers

If sampled continuously, a GPS receiver may quickly drain a battery on a mobile phone [17]. An obvious strategy to reduce the energy consumption is to periodically sample ("duty cycle") the receiver. The energy consumption of a periodically sampled GPS receiver depends on the time needed to obtain a location fix, which depends on the sampling interval. We carried out a number of experiments to measure the energy consumption of continuous and periodic GPS sampling. For the measurements, we used HTC Desire smartphones running Android v2.2.

First, we measured the energy consumption of continuous GPS sampling under various satellite visibility conditions. The consumption is measured based on the electric current drained from the battery, whose milliampere value is obtained from the battery driver. In each test, we calculated the average power draw over a one-hour period. Regardless of the satellite visibility, this average power draw $P_{GPS}$ was close to 305 mW (~78 mA @ 3.9 V) on top of the measured base consumption of the phone. This is consistent with the 80 to 85 mA measured on Android phones in [23] and somewhat less than 370 mW on Symbian Nokia N95 phone in [17].

Second, we measured the time-to-fix (TTF) for various GPS sampling intervals. Before each test, up-to-date satellite almanac and ephemeris data was downloaded using Wi-Fi connection. Hence, whenever sampled, phones' GPS receivers performed so-called "hot start" with valid satellite information. After a hot start,

TTF depends on how fast a GPS receiver can tune to the carrier frequencies of visible satellites and synchronize with their signals. The carrier frequencies are constantly shifted due to the Doppler effect. In environments with lots of shadowing and multipath scattering, it may take a few tens of seconds to acquire a fix. The measurements were performed in one of the Disney's theme parks, where phones were carried at the walking speed of a typical theme park visitor. TTFs were measured for four sampling intervals $T_S$ (30, 60, 120, and 300 seconds). The sampling interval is the time elapsed since the last fix was obtained until the next one is requested. The TTFs for each sample, as well as the average TTFs for each of the sampling intervals are shown in Fig 1. Based on the results, we construct a model that describes the average TTF as a function of the sampling interval $T_S$:

$$\mathrm{TTF}(T_S) = \begin{cases} 10, & T_S \leq 30 \\ 10 + 8 \cdot \log_{10}(T_S/30), & 30 < T_S < 120 \\ 15, & T_S \geq 120 \end{cases} \quad (1)$$

The function $\mathrm{TTF}(T_S)$ is also plotted in Fig. 1. The model (1) assumes typical pedestrian walking speed. We performed another set of measurements where phones were static. As expected, TTFs were significantly shorter because Doppler and multipath effects were less prominent. Based on the results, which we omit for brevity, we construct a model that describes the average TTF when phones are static:

$$\mathrm{TTF}(T_S) = \begin{cases} 4, & T_S \leq 30 \\ 4 + 2 \cdot \log_{10}(T_S/30), & 30 < T_S < 300. \\ 6, & T_S \geq 300 \end{cases} \quad (2)$$

The energy consumed to acquire a fix is then given by $E_{FIX}(T_S) = P_{GPS} \cdot TTF(T_S)$. Although models in (1) and (2) are not thoroughly validated (i.e. by comparing results from various theme parks), we believe that they better reflects the reality of theme park scenarios than models described in literature. For example, the model in [18] assumes a constant TTF of six seconds for sampling intervals longer than 30 seconds.

## 3.2 Energy Consumption of Wireless Interfaces

Collaborative GPS localization relies on ad hoc communication between neighboring devices using their wireless (e.g. Wi-Fi) interfaces. This communication incurs energy overhead. The overhead includes the energy spent to transmit/receive location information to/from the neighbors. We assume that the energy spent for idle listening is not a part of this overhead because, in ad

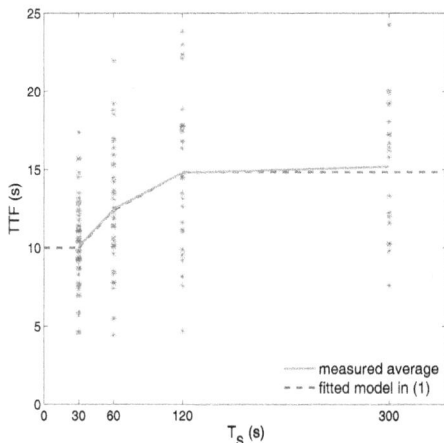

**Figure 1. TTFs measured for various sampling intervals on phones carried by theme park visitors.**

hoc networks, devices anyway need to listen for incoming traffic from their neighbors. In the following, we assume that devices are equipped with Wi-Fi interfaces, which are nowadays available on most phones. To calculate the energy overhead in our simulations, we adopt the per-packet energy consumption model introduced and validated in [24]. The model assumes that the energy spent on top of idle listening ($energy^+$) to send or receive a packet in ad hoc mode is given by

$$energy^+ = m \times size + b,$$

where $size$ is measured in bytes. Hence, the consumption has a fixed component associated with the power state changes and channel acquisition, and an incremental component, which is proportional to the size of the packet. The values of coefficients $m$ and $b$ depend on whether a packet is broadcasted or sent point-to-point. Table 1 summarizes the values of coefficients $m$ and $b$ measured in [24]. As an example, in the second-to-last column of the table, we calculate the energy consumption (on top of idle listening) to send/receive a 100-byte packet. The total energy consumption, assuming idle listening power of 741 mW, is given in the last column of the table. It is in the order of a few hundreds of microjoules. We use the model in our simulations to make a rough estimate of the energy consumed by a Wi-Fi interface whenever it transmits or receives one of the protocol messages of the collaborative GPS localization protocol. In practice, the consumption depends on particular Wi-Fi chipset, transmission rate, and other factors that we do not consider: The goal is to indicate general orders of magnitude.

## 4. POSITION ERROR MODEL

Let us first assume that a mobile handheld device is sampling its GPS receiver to position itself without help from neighboring devices. The device maintains its position estimate $p = (x, y)$, which is updated with each new GPS sample. Our position error model for this scenario takes into account two sources of errors: position uncertainty of the last GPS update and the distance traveled since the update. Let $\tau$ be the time elapsed since the last position update ($\tau$ is the age of the position estimate $p$). Let $e(\tau)$ be the position error of $p$ with respect to the current true position $P(\tau) = (X(\tau), Y(\tau))$:

$$e(\tau) = |P(\tau) - p| = \sqrt{(X(\tau) - x)^2 + (Y(\tau) - y)^2}$$

The model assumes that the expected position error $\mathrm{E}[e(\tau)]$, which measures the uncertainty of position $p$ is given by

$$\mathrm{E}[e(\tau)] = \mathrm{E}[e(0)] + \tau \cdot \overline{v(\tau)}. \quad (3)$$

where $\mathrm{E}[e(0)] = \mathrm{E}[e_{GPS}]$ is the expected horizontal position error of the GPS update and $\overline{v(\tau)}$ is the estimated average speed of the phone during the period $(0, \tau)$. Ways to estimate $\mathrm{E}[e_{GPS}]$ and $\overline{v(\tau)}$ are discussed in the following. Note that (3) may overestimate the actual error since the displacement of the phone (with respect to its position at $\tau = 0$) depends on the movement trajectory and it is often smaller than $\tau \cdot \overline{v(\tau)}$.

**Table 1. Wi-Fi energy consumption measurements in [24] assume data rate of 11 Mb/s. Idle listening power is 741 mW.**

| | m (μJ/byte) | b (μJ) | energy$^+$ 100 bytes (μJ) | tot. energy 100 bytes (μJ) |
|---|---|---|---|---|
| point-to-point send | 0.48 | 431 | 479 | 533 |
| broadcast send | 2.1 | 272 | 482 | 536 |
| point-to-point receive | 0.12 | 316 | 328 | 382 |
| broadcast receive | 0.26 | 50 | 76 | 130 |

## 4.1 Position Error of GPS

The horizontal position error of GPS ($e_{GPS}$) depends on many factors: number and constellation of visible satellites, satellite clock and ephemeris data errors, atmospheric propagation delay, multipath fading, and GPS receiver quality. It can be written as

$$e_{GPS} = \sqrt{e_{xGPS}^2 + e_{yGPS}^2}$$

where $e_{xGPS}$ and $e_{yGPS}$ are random variables that correspond to the errors in $x$ (longitude) and $y$ (latitude) directions, respectively. The two random variables are often assumed to follow a normal distribution with zero mean, which has been confirmed in [25]. If we assume the same variances in both directions, then the horizontal error $e_{GPS}$ is Rayleigh distributed:

$$P(e_{GPS} \leq \varepsilon) = 1 - e^{-\frac{e^2}{2\sigma^2}} = 1 - e^{-\frac{e^2}{\sigma_{rms}^2}},$$

where $\sigma_{rms} = 2/\sqrt{\pi} \cdot E[e_{GPS}]$ is the root-mean-squared (RMS) error. When $\varepsilon = \sigma_{rms}$, then $P(e_{GPS} \leq \varepsilon) = 1 - e^{-1} \approx 0.63$. Hence, 63% of errors fall within a circle of radius $\sigma_{rms}$. Therefore, the RMS error $\sigma_{rms}$ is often referred to as "63% error distance" and denoted by $dRMS$. This is the value what most GPS receivers estimate and report as their accuracy. Some receivers report "95% error distance" denoted by $2dRMS$. There is a simple relationship between the two measures: $2dRMS \approx 1.73 \cdot \sigma_{rms} \approx 1.95 \cdot E[e_{GPS}]$. Hence, $E[e_{GPS}]$ can be easily estimated from the accuracy reported by GPS receivers. In our simulations, $E[e_{GPS}]$ is drawn from a GPS dataset, which is described in Section 6.

## 4.2 Speed Estimation

Accurate estimation of $\overline{v(\tau)}$ on a phone is not a trivial problem. One option would be to assume that $\overline{v(\tau)}$ is equal to the speed reported by the GPS receiver at $\tau = 0$. GPS receivers typically estimate the speed from their position samples and from the measured Doppler shift of satellite signals. It has been shown, however, that these estimates are very unreliable at pedestrian speeds [18]. Another possibility is to adopt an upper bound on the walking speed, which may be $v_{max} = 1.5$ m/s for a vigorous park visitor, and assume that $\overline{v(\tau)} = v_{max}$. Clearly, this may grossly overestimate the expected position error $E[e(\tau)]$ in (3). Inertial sensors on the phone, such as the accelerometer, can be used to refine the estimate. Readings from the accelerometer can be interpreted (e.g. based on a threshold crossing) as a binary indicator if a person is moving or not. The speed $\overline{v(\tau)}$ is then estimated as a product of $v_{max}$ and the average value of the binary indicator over the period $(0, \tau)$. Evaluation of different methods to estimate $\overline{v(\tau)}$ is out of the scope of this paper. In our simulations, we assume that $\overline{v(\tau)}$ is estimated from the distance between the last two position updates and the time elapsed between them. The drawback of this method is that positioning errors introduce errors in the speed estimate. We put a lower cap on $\overline{v(\tau)}$ to 0.2 m/s. Otherwise, location updating would stall when $\overline{v(\tau)} = 0$.

## 5. COLLABORATIVE LOCALIZATION

Our collaborative localization protocol works as follows: Each mobile phone $i$ maintains an up-to-date position error estimate $E[e_i(\tau_i)]$ according to (3). If its error estimate exceeds the maximum tolerable position error $e_{max}$, the phone $i = 0$ broadcasts a *location update request* using its wireless interface. The request contains the phone's last position estimate $p_0 = (x_0, y_0)$ and position error estimate $E[e_0(\tau_0)]$. Every neighbor $i$ within the transmission range $r$ who receives the request will estimate its distance $d_i$ to the sender and compare the sender's error estimate

$E[e_0(\tau_0)]$ to its own error estimate $E[e_i(\tau_i)]$. If $E[e_i(\tau_i)] + d_i < E[e_0(\tau_0)]$, the neighbor $i$ will send a *location update response* with its position and position error estimates $p_i$ and $E[e_i(\tau_i)]$. Hence, a neighbor will respond to the request if it has a more accurate position estimate that the originator of the request, taking into account the distance between the two. The distance $d_i$ can be estimated using received signal strength (RSS) or time of flight (TOF) based ranging techniques. RSS ranging is supported on commodity phones, but it suffers from low accuracy. TOF ranging may provide more accurate distance estimation, but typically requires changes to the phones' hardware and/or protocol stack. A software-based TOF ranging technique for Wi-Fi is described in [26]. Unfortunately, it can only be implemented on phones with reconfigurable open-source Wi-Fi drivers, which are presently rare. We focus on scenarios where the transmission range $r$ is relatively small compared to the target location accuracy $e_{max}$. In such scenarios, ranging capabilities are not essential. In our protocol, neighboring phones assume that $d_i = r$, which accounts for the worst-case scenario.

Assume now that the phone obtains *location update responses* from $N$ neighbors:

- If $N = 0$, the phone triggers its GPS receiver and waits for a position fix. After it obtains a fix, the phone updates its position estimate $p_0$, resets the age of the position estimate to $\tau_0 = 0$, calculates the expected position error $E[e_0(0)]$ according to (3), and broadcasts a *location update notification* containing $p_0$ and $E[e_0(0)]$ to its current neighbors.

- If $N > 0$, the responses from multiple neighbors are combined in order to update the phone's position estimate $p_0$. We consider the following two combining schemes:

*Substitution:* Position estimate $p_0$ is substituted with $p_k$, where

$$k = \arg \min_{1 \leq i \leq N} (E[e_i(\tau_i)] + r)$$

Hence, among the neighbors' positions $p_i$, $1 \leq i \leq N$, the one with the smallest expected error is used as a new position estimate $p_0$. Correspondingly, the new position error is

$$E[e_0(0)] = E[e_k(\tau_k)] + r.$$

*Averaging:* position $p_0$ is calculated as a weighted average of $p_i(x_i, y_i), 1 \leq i \leq N$:

$$p_0 = \left( \sum_{i=1}^N w_i x_i, \sum_{i=1}^N w_i y_i \right), \quad (4)$$

where weights $w_i$ are chosen so to minimize the expected squared error $\sigma^2$ of the position estimate $p_0$ with respect to the current true position $P_0(\tau_0)$:

$$\sigma^2 = E[|P_0(\tau_0) - p_0|^2] =$$
$$= E\left[ \left( X_0(\tau_0) - \sum_{i=1}^N w_i x_i \right)^2 + \left( Y_0(\tau_0) - \sum_{i=1}^N w_i y_i \right)^2 \right].$$

Assuming that $(X_0(\tau_0) - x_i, Y_0(\tau_0) - y_i)$, $1 \leq i \leq N$ are uncorrelated random variables, it is easy to show that $\sigma^2$ is

$$\sigma^2 = \sum_{i=1}^N w_i^2 \sigma_i^2, \quad (5)$$

where $\sigma_i^2 = E[|P_0(\tau_0) - p_i|^2] = E[(X_0(\tau_0) - x_i)^2 + (Y_0(\tau_0) - y_i)^2]$ is the expected squared error of $p_i$ with respect to $P_0(\tau_0)$. From (5), we obtain that $\sigma^2$ is minimized for

$$w_i = \frac{1/\sigma_i^2}{\sum_{k=1}^N 1/\sigma_k^2} \quad (6)$$

However, $w_i$ cannot be calculated from (6) because $\sigma_i^2$, $1 \leq i \leq N$ are unknown. In order to express $w_i$ in terms of $E[e_i(\tau_i)]$, we use the following approximations: Since $|P_0(\tau_0) - p_i| = |P_0(\tau_0) - P_i(\tau_i) + e_i(\tau_i)|$ and $|P_0(\tau_0) - P_i(\tau_i)| \approx r$, it follows

$$\sigma_i^2 \approx \mathrm{E}[(e_i(\tau_i) + r)^2] =$$

$$= \mathrm{E}[(e_i(\tau_i) + r - \mathrm{E}[e_i(\tau_i) + r])^2] + \mathrm{E}[e_i(\tau_i) + r]^2 =$$

$$= \mathrm{E}[e_i(\tau_i) + r]^2 \cdot \left( \frac{\mathrm{E}[(e_i(\tau_i) + r - \mathrm{E}[e_i(\tau_i) + r])^2]}{\mathrm{E}[e_i(\tau_i) + r]^2} + 1 \right) =$$

$$= (\mathrm{E}[e_i(\tau_i)] + r)^2 \cdot (\gamma_i^2 + 1)$$

where $\gamma_i$ is the ratio of the standard deviation and the mean of $e_i(\tau_i) + r$. Given that $e_i(\tau_i)$, $1 \le i \le N$ follow the same distribution, their $\gamma_i$ ratios are equal ($\gamma_i = \gamma$). By substituting $\sigma_i^2$ in (6), we obtain

$$w_i = \frac{\frac{1}{(\mathrm{E}[e_i(\tau_i)] + r)^2}}{\sum_{k=1}^{N} \frac{1}{(\mathrm{E}[e_k(\tau_k)] + r)^2}}$$

Now the weighted average position $p_0$ can be calculated by substituting $w_i$ in (4). Starting from (5) and following the same reasoning, it can be shown that the expected error $\mathrm{E}[e_0(0)]$ is

$$\mathrm{E}[e_0(0)] = \sqrt{\sum_{i=1}^{N} w_i^2 (\mathrm{E}[e_i(\tau_i)] + r)^2}.$$

Once $p_0$ and $\mathrm{E}[e_0(0)]$ are determined, either by substitution or by averaging, a *location update notification* is broadcasted to the neighbors, which in turn update their position estimates $p_i$ if $\mathrm{E}[e_0(0)] + r < E[e_i(\tau_i)]$.

# 6. EVALUATION SETUP

We analyze and compare the energy consumptions of non-collaborative and collaborative GPS localization of theme park visitors using simulations. The scenario assumes that visitors' phones maintain estimates of their current position errors $\mathrm{E}[e(\tau_i)]$ defined in (3) and, when the error exceeds $e_{max}$, they either trigger their GPS receiver (in case of non-collaborative localization) or execute the protocol described in Section 5 (in case of collaborative localization). We measure the number of GPS activations, number of protocol messages exchanged between the phones, and the deviation of the position estimates from the ground truth positions.

The performance of collaborative localization depends strongly on the density and mobility of the visitors/phones. The density affects the number of potential collaborators. The mobility affects the position errors $\mathrm{E}[e(\tau_i)]$ and the frequency of collaboration opportunities. In our simulations, density and mobility of visitors are driven by real-world data: As a part of a research study unrelated to this paper, we collected 910 GPS traces by handing out GPS-enabled phones to the visitors of the Epcot theme park in Florida [5]. The layout of the park is shown in Fig. 2 (left). The phones were handed out between 8am and 1pm at the entrance gate, and collected when the visitors were exiting the park. The spatial distribution of the phones at different times of the day is shown in Fig. 2 (right). The bell-shaped curve in Fig. 3 shows the number of phones in the park as a function of time. The phones were sampling their GPS receivers on average every two minutes. In addition to the geo-coordinates, GPS accuracy was also logged. We discarded waypoints whose accuracy was worse than 25 m. We also discarded tracks shorter than two hours or containing less than 50 waypoints. Results presented in the following sections are based on the remaining 825 out of 910 tracks. We interpolated the movements of visitors between the waypoints assuming straight-line movements. The interpolated trajectories are used as "ground truth" location data to drive the mobility of nodes in our simulations. The original dataset, which includes waypoints with accuracy is worse than 25 m, is used to generate the expected horizontal position error $\mathrm{E}[e_{GPS}]$ of the simulated GPS receivers.

Whenever a simulated node triggers its GPS receiver, the time-to-fix (TTF) and the expected error of the fix $\mathrm{E}[e_{GPS}]$ are generated: TTF is generated based on the time $T_S$ elapsed since the receiver was sampled last time and the visitor's average *ground-truth* walking speed during that time: If the speed is higher than 0.2 m/s, TTF is generated according to (1). Otherwise, it is generated according to (2). The energy consumption of the GPS receiver $E_{GPS}$ increases by $E_{FIX}(T_S) = P_{GPS} \cdot TTF(T_S)$ with every new sample. $\mathrm{E}[e_{GPS}]$ is drawn from the empirical distribution observed in the traces. The average 95 % error distance in the traces is $\overline{2dRMS} = 21.3$ m, which corresponds to the average $\overline{\mathrm{E}[e_{GPS}]} = \overline{2dRMS}/1.95 \approx 11$ m. The phones were equipped with GPS chipsets based on Qualcomm's gpsOne technology, which is world's most widely used GPS technology for handsets [27].

We also keep track of the number of protocol messages (location update requests, location update responses, and location update notifications) exchanged between the nodes. In order to estimate the energy overhead of a Wi-Fi interface $E_{WiFi}$, we introduce the following simplifications:

• The size of each protocol message is 100 bytes at the physical layer. This is more than sufficient to accommodate

**Figure 2. Left: The layout of the Epcot theme park in Florida. Right: spatial distribution of park visitors at 11am, 13pm, 16:30pm, and 17:30pm (clockwise starting from the top left image).**

49

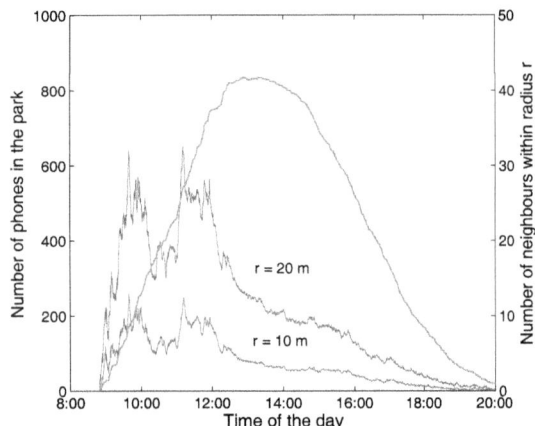

**Figure 3. Number of phones and the average number of neighbors within radius $r$ at different times of the day.**

location information payload (position estimate $p_i$ and position error estimate $E[e_i(\tau_i)]$) and all upper-layer headers.

- Sending or receiving a protocol message consumes 600 μJ of energy, regardless if whether the message is sent/received in broadcast mode (location update request, location update notification) or unicast mode (location update response). According to Table 1, this overestimates the actual consumption of Wi-Fi interfaces.

- The energy consumed to send a protocol message does not depend on the transmission range $r$. It has been shown in [28] that transmit power has a minor impact on the energy consumption of Wi-Fi interfaces.

Therefore, the energy consumption of a Wi-Fi interface $E_{WiFi}$ increases by 600 μJ whenever it sends or receives a protocol message. The total energy consumption is given by the sum of $E_{GPS}$ and $E_{WiFi}$ for the collaborative GPS localization, and by $E_{GPS}$ only for the non-collaborative GPS localization.

## 7. PERFORMANCE RESULTS

We present the performance results for non-collaborative and collaborative GPS localization (with position substitution and averaging) for various values of the maximum tolerable position error $e_{max}$ and transmission range $r$. Each simulated node, whose arrival/departure and mobility in the park are driven by one of the ground-truth traces, maintains the following statistics: number of GPS activations, energy consumed by the GPS receiver, number of sent/received protocol messages, energy consumed by the Wi-Fi interface, and deviation of the position estimate from the ground-truth position. At the end of each simulation run, we calculate for each node the average number of GPS activation and protocol messages per hour, average GPS and Wi-Fi energy consumption per hour, and average deviation from the ground-truth position. These values are then averaged over all devices and shown in Table 2.

In case of non-collaborative localization, the number of GPS activations decreases from 59.4/h to 12.4/h as $e_{max}$ increases from 25 m to 75 m, and so does the energy consumption, although the energy per activation increases (3.6 J/activation for $e_{max} = 25$ m vs. 4.5 J/activation for $e_{max} = 75$ m). This is because TTF increases when GPS is sampled less frequently. The average consumption of GPS receivers for $e_{max} = 25$ m is 213.6 J/h. For a comparison, a smartphone in the suspended state consumes ~100 J/h (HTC Dream: 26.6 mW or 95.6 J/h, Google Nexus One: 24.9 mW or 89.6 J/h) [29]. When $e_{max}$ increases to 75 m, the GPS consumption drops to 56 J/h, which is however still significant. Surely, this consumption is dwarfed by the idle listening consumption of Wi-Fi, but this is likely to change (e.g. with the use of Power Save Mode and Opportunistic Power Save protocol in Wi-Fi Direct). Table 2 also shows that the average deviation from the ground-truth position is well below $e_{max}$.

Collaborative localization based on position substitution with $r = 10$ m reduces the number of GPS activations and, therefore, the GPS energy consumption. The reduction depends on $e_{max}$. For $e_{max} = 25$ m, the GPS consumption is 151 J/h, which is 71% of 213 J/h consumed by non-collaborative localization. For $e_{max} = 75$ m , it decreases to 24.3 J/h, which is 43% of 56 J/h consumed by non-collaborative localization. The energy overhead of Wi-Fi is negligible (< 1 J/h). The average deviation from the ground-truth position is almost the same as with non-collaborative localization. The results show that collaboration significantly reduces the energy consumption, especially when the maximum

**Table 2. Performance results for non-collaborative and collaborative GPS localization for various values of $e_{max}$ and $r$. The results are averaged over time spent in the park and over all devices.**

| GPS localization method | $r$ (m) | $e_{max}$ (m) | GPS activations (1/h) | GPS energy (J/h) | Protocol messages (1/h) | Wi-Fi energy (J/h) | GPS+Wi-Fi energy (J/h) | Deviation (m) |
|---|---|---|---|---|---|---|---|---|
| non-collaborative | n.a. | 25 | 59.4 | 213.6 | n.a. | n.a. | 213.6 | 9.9 |
| | | 50 | 21.7 | 92.3 | n.a. | n.a. | 92.3 | 16.4 |
| | | 75 | 12.4 | 56.0 | n.a. | n.a. | 56.0 | 23.0 |
| collaborative (substitution) | 10 | 25 | 43.5 | 151.1 | 983.4 | 0.6 | 151.7 | 9.8 |
| | | 50 | 11.6 | 44.8 | 323.3 | 0.2 | 45.0 | 16.5 |
| | | 75 | 5.9 | 24.3 | 176.6 | 0.1 | 24.4 | 23.6 |
| | 20 | 25 | 64.6 | 220.9 | 4300.8 | 2.6 | 223.5 | 10.3 |
| | | 50 | 11.7 | 42.0 | 866.0 | 0.5 | 42.5 | 16.9 |
| | | 75 | 4.5 | 17.3 | 515.9 | 0.3 | 17.6 | 24.3 |
| collaborative (averaging) | 10 | 25 | 41.0 | 143.8 | 764.3 | 0.5 | 144.3 | 9.8 |
| | | 50 | 10.2 | 40.4 | 196.2 | 0.1 | 40.5 | 16.8 |
| | | 75 | 5.0 | 20.9 | 111.8 | 0.1 | 21.0 | 25.3 |
| | 20 | 25 | 59.4 | 205.6 | 3608.4 | 2.2 | 207.8 | 10.2 |
| | | 50 | 9.5 | 36.2 | 490.6 | 0.3 | 36.5 | 17.6 |
| | | 75 | 3.1 | 12.6 | 201.1 | 0.1 | 12.7 | 26.2 |

tolerable error increases. With an optimal selection of the transmission range, the energy consumption can be further reduced. The optimal selection depends on an the interplay between $r$ and $e_{max}$: On one hand, a larger range increases the number of neighbors and, therefore, the chance to obtain a location estimate without GPS. On the other hand, the accuracy of position estimates obtained from the neighbors decreases (since every neighbor is assumed to be $r$ meters away) and might not be sufficient to suppress GPS activations if $r$ approaches $e_{max}$. The results for $r = 20$ m show that collaboration actually increases the energy consumption when $e_{max} = 25$ m. However, when $e_{max} = 75$ m, the consumption is only 31% of 56 J/h consumed by non-collaborative localization, down from 43% with $r = 10$ m. Results achieved with the position averaging algorithm are also shown in Table 2. This algorithm utilizes location information from multiple neighbors to obtain more accurate position estimates. Therefore, it further reduces the number of GPS activations, hence, the energy consumption. The results from Table 2 are summarized in Fig. 4. The figure shows that 1) the energy savings of collaborative localization increase with $e_{max}$, 2) the averaging algorithm provides additional savings compared to the substitution algorithm, 3) the optimal transmission range $r$ depends on $e_{max}$, and 4) even for ranges that are marginally smaller than $e_{max}$ (e.g. $r = 20$ m, $e_{max} = 25$ m) collaborative localization may provide energy savings without distance estimation between neighbors.

The results presented so far show the time-average performance for a random device for the entire duration of its stay in the park. A typical visit to the Epcot lasts 5-6 hour. During this time, the number of visitors and their spatial distribution change, and so does the number of neighbors within the radius $r$, as shown in Fig. 3. Notice that the number of neighbors is weakly correlated with the number of devices in the park because visitors tend to gather at certain locations in the park at certain times of the day. For example, in the morning hours, they crowd in the front section of the park, as shown in Fig. 2 (right). Later they may gather to watch a street performance or similar event. Once they spread across the park area in the late afternoon, the correlation becomes stronger. To show the impact of the number of neighbors on the performance of collaborative localization, we observe three one-hour periods listed in Table 3 when the number of neighbors is relatively constant. Energy consumption of collaborative localization based on position averaging for the three periods and various $e_{max}$ is shown in Fig. 5 as a percentage of energy

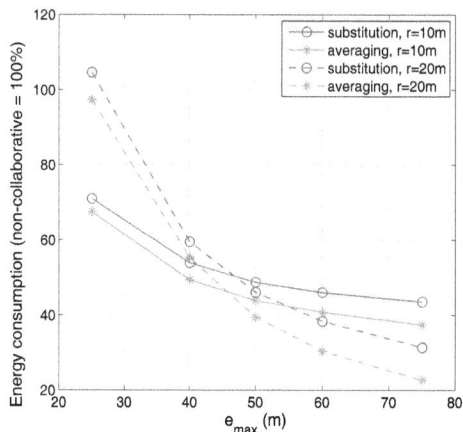

Figure 4. Energy consumption of collaborative GPS localization for various $e_{max}$.

consumed by non-collaborative localization. As expected, the consumption decreases with the number of neighbors. The decrease is not uniform: When the number of neighbors within 10 m (20 m) increases beyond 6.5 (18.9), which roughly corresponds to the density of 1.5-2 devices per 100 m², the consumption decreases only marginally. Interestingly, the results in Fig. 5 suggest that optimal $r$ does not depend on device density. Clearly this is not true in general because optimal $r$ tends to zero when device density tends to infinity. It, however, implies that, for scenarios of interest, close-to-optimal $r$ (fine-tuning of $r$ is anyway not possible) can be chosen solely based on $e_{max}$. Based on Figs. 3 and 4, it appears that $0.2e_{max} < r < 0.4e_{max}$ is a good rule of thumb for choosing the range in the considered scenario.

## 8. CONCLUSION

We evaluated two collaborative GPS localization protocols based on position substitution and position averaging. The evaluation is based on realistic simulations where the mobility of people is driven by real-world data from a theme park. Since the energy overhead of wireless interfaces is negligible in ad hoc networks where devices need to listen for incoming traffic anyway, the proposed schemes can provide significant energy savings compared to the non-collaborative GPS localization. One parameter that can be engineered to maximize the energy savings is the transmission range. In the absence of distance estimation between neighbors, the optimal range depends heavily on the maximum tolerable position error. We provide guidelines for choosing the range in the considered scenario. If distance estimation would be available on commodity phones, the consumption would always decrease with the range as long as the transmission power is a negligible part of that consumption.

The important problem of security (i.e. position information integrity) has not been addressed in this work. Without complex trust/reputation schemes, the collaborative localization protocols are prone to malicious announcements of incorrect positions. However, a coordinated effort of a significant number of malicious users is needed to introduce a persistent positioning error. We have not considered use case scenarios for collaborative GPS localization that would attract such coordinated attacks.

## 9. REFERENCES

[1] Wi-Fi Alliance: Wi-Fi Direct. [Online]. http://www.wi-fi.org/Wi-Fi_Direct.php. [Accessed: Oct. 30, 2011].

[2] Bluetooth Low Energy. [Online]. http://www.bluetooth.com/Pages/Low-Energy.aspx. [Accessed: Oct. 30, 2011].

[3] Nokia Instant Community gets you social. [Online]. http://conversations.nokia.com/2010/05/25/nokia-instant-community-gets-you-social/. [Accessed: Oct. 30, 2011].

[4] D. M.-K. Low et al., "Group formation using anonymous broadcast information," U.S. Patent 233 358, 2010.

[5] V. Vukadinovic and S. Mangold, "Opportunistic Wireless

Table 3. Average number of neighbors at different periods of the day.

| Period | Neighbors (r = 10m) | Neighbors (r = 20m) |
|---|---|---|
| 14h - 15h | 2.8 | 9.3 |
| 10h - 11h | 6.5 | 18.9 |
| 11h - 12h | 9.1 | 25.8 |

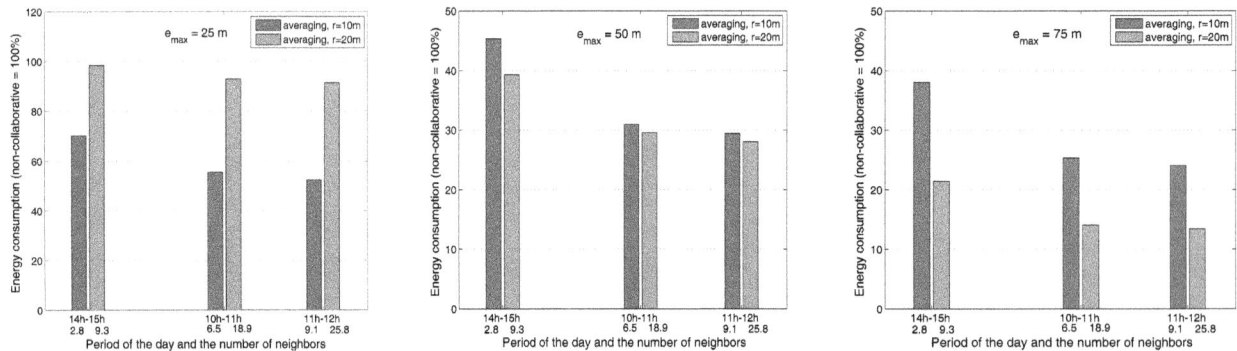

**Figure 5. Energy consumption of collaborative localization depends on the average number of neighbors at different periods of the day.**

Communication in Theme Parks: A Study of Visitors Mobility," *Proc. ACM MobiCom Workshop on Challenged Networks (CHANTS)*, Las Vegas, USA, 2011.

[6] H. Wymeersch, J. Lien, and M. Z. Win, "Cooperative localization in wireless networks," *Proc. IEEE*, vol. 97, no. 2, pp. 427–450, Feb. 2009.

[7] R. Kurazume, S. Nagata, and S. Hirose, "Cooperative positioning with multiple robots," *Proc. IEEE Int. Conf. Robotics and Automation*, San Diego, USA, 1994.

[8] A. I. Mourikis and S. I. Roumeliotis, "Performance analysis of multirobot cooperative localization," *IEEE Trans. Robotics*, vol. 22, no. 4, pp. 666–681, 2006.

[9] P. Barooah, W. J. Russell, and J. P. Hespanha, "Approximate distributed Kalman filtering for cooperative multi-agent localization," *Proc. Int. Conf. Distributed Computing in Sensor Systems*, Santa Barbara, USA, 2010.

[10] P. Zhang and M. Martonosi, "LOCALE: Collaborative Localization Estimation for Sparse Mobile Sensor Networks," *Proc. Int. Conf. Information Processing in Sensor Networks*, St. Louis, USA, 2008.

[11] R. Jurdak, P. Corke, D. Dharman, and G. Salagnac, "Adaptive GPS Duty Cycling and Radio Ranging for Energy-Efficient Localization," *Proc. ACM Sensys*, Zurich, Switzerland, 2010.

[12] F. Zorzi et al., "Exploiting opportunistic interactions for localization in heterogeneous wireless systems," *Proc. NEWCOM++ / COST 2100 Joint Workshop on Wireless Communications*, Paris, France, 2011.

[13] L. Chan et al., "Collaborative Localization: Enhancing WiFi-Based Position Estimation with Neighborhood Links in Clusters," *Proc. Int. Conf. Pervasive Computing*, Dublin, Ireland, 2006.

[14] F. Della Rosa et al., "Experimental Activity on Cooperative Mobile Positioning in Indoor Environments," *Proc. IEEE Worshop on Advanced Experimental Activities on Wireless Networks & Systems*, Helsinki, Finland, 2007.

[15] C. Mensing et al., "Performance Assessment of Cooperative Positioning Techniques", *Proc. ICT Future Networks and Mobile Summit*, Florence, Italy, 2010.

[16] J. Hemmes, D. Thain, and C. Poellabauer, "Cooperative Localization in GPS-Limited Urban Environments," *Proc. Int. Conf. Ad Hoc Networks (AdHocNets)*, Niagara Falls, Canada, 2009.

[17] J. Paek, J. Kim, and R. Govindan, "Energy-efficient rate-adaptive gps-based positioning for smartphones," *Proc. ACM MobiSys*, San Francisco, USA, 2010.

[18] M. B. Kjaergaard, J. Langdal, T. Godsk, and T. Toftkjaer, "Entracked: energy-efficient robust position tracking for mobile devices," *Proc. ACM MobiSys*, Krakow, Poland, 2009.

[19] Z. Zhuang, K.-H. Kim, and J. P. Singh, "Improving energy efficiency of location sensing on smartphones," *Proc. ACM MobiSys*, San Francisco, USA, 2010.

[20] F. Ben Abdesslem, A. Phillips, and T. Henderson, "Less is more: energy-efficient mobile sensing with SenseLess," *Proc. ACM MobiHeld*, Barcelona, Spain, 2009.

[21] D. H. Kim, Y. Kim, D. Estrin, and M. B. Srivastava, "SensLoc: sensing everyday places and paths using less energy," *Proc. ACM SenSys*, Zurich, Switzerland, 2010.

[22] I. Constandache, S. Gaonkar, M. Sayler, R.R. Choudhury, and L. Cox, "EnLoc: energy-efficient localization for mobile phones," *Proc. IEEE INFOCOM Mini Conference*, Rio de Janeiro, Brazil, 2009.

[23] J. Sharkey, "Coding for life – battery life, that is," *Google IO Developer Conference*, San Francisco, USA, 2009.

[24] L. M. Feeney and M. Nilsson, "Investigating the energy consumption of a wireless network interface in an ad hoc networking environment," *Proc. IEEE Infocom,* Anchorage, USA, 2001.

[25] P. A. Zandbergen, "Positional accuracy of spatial data: Non-normal distributions and a critique of the National Standard for Spatial Data Accuracy," *Transactions in GIS*, vol 12, pp. 103–130, 2008.

[26] D. Giustiniano and S. Mangold, "Demo: Distance Tracking Using WLAN Time of Flight," *Proc. ACM MobiSys*, Washington, USA, 2011.

[27] Navigating LBS Platform. [Online]. http://developer.qualcomm.com/file/617/navigatinglbsplatformspdf. Qualcomm Inc., 2010. [Accessed May 17, 2011].

[28] D. Halperin, B. Greensteiny, A. Shethy, and D. Wetherall, "Demystifying 802.11n power consumption," *Proc. USENIX HotPower*, Vancouver, Canada, 2010.

[29] A. Carroll and G. Heiser, "An analysis of power consumption in a smartphone," *Proc. USENIX Annual Technical Conference*, Boston, USA, 2010.

# Making the Most of Your Contacts: Transfer Ordering in Data-Centric Opportunistic Networks

Christian Rohner
Uppsala University, Sweden
christian.rohner@it.uu.se

Fredrik Bjurefors
Uppsala University, Sweden
fredrik.bjurefors@it.uu.se

Per Gunningberg
Uppsala University, Sweden
per.gunningberg@it.uu.se

Liam McNamara
Uppsala University, Sweden
liam.mcnamara@it.uu.se

Erik Nordström
Princeton University, USA
enordstr@cs.princeton.edu

## ABSTRACT

Opportunistic networks use unpredictable and time-limited contacts to disseminate data. Therefore, it is important that protocols transfer useful data when contacts do occur. Specifically, in a data-centric network, nodes benefit from receiving data *relevant* to their interests. To this end, we study five strategies to *select and order* the data to be exchanged during a limited contact, and measure their ability to promptly and efficiently deliver highly relevant data.

Our trace-driven experiments on an emulation testbed suggest that nodes benefit in the short-term from ordering data transfers to satisfy local interests. However, this can lead to suboptimal long-term system performance. Restricting sharing based on matching nodes' interests can lead to segregation of the network, and limit useful dissemination of data. A non-local understanding of other nodes' interests is necessary to effectively move data across the network. If ordering of transfers for data relevance is not explicitly considered performance is comparable to random, which limits the delivery of individually relevant data.

## Categories and Subject Descriptors

C.2.0 [**Computer-Communication Networks**]: General—Data communications

## Keywords

Opportunistic Networking, Data-Centric, Relevance

## 1. INTRODUCTION

In opportunistic networks, mobile nodes exchange data when they are within wireless communication range. Data items are disseminated via neighbours to enable them to reach interested parties, with nodes using the *store-carry-forward* principle to collaboratively establish a data relaying service.

Due to the unpredictability of node contact duration it is not known how many successful item transfers will be performed during a contact. Therefore, a data dissemination protocol must judi-

ciously decide *what* data to exchange and in *which order*, to make the best use of a time-limited contact and to achieve the best "value" for the network at minimal cost.

Transferring an item of data represents a selection to spread that particular item instead of another. The receiver then gains the utility from the transferred items and can also spread it to other nodes in the network. When choosing a dissemination system, a key consideration is balancing local benefit to the receiver and potential global benefit for the entire system.

For the rest of the paper, we shall use the following terminology to refer to two aspects of opportunistic data dissemination. Deciding whether an item should be transferred to a given host is a *selection* decision, deciding which order the selected items should be sent in is called an *ordering* decision.

Numerous selection algorithms have been proposed to efficiently choose nodes that will carry data to other nodes in the network [10, 12, 18, 21, 26]. These decisions are often made based on node properties, for example, contact history or role in community structures. Deciding which items to drop from a finite data item buffer has been extensively studied, with strategies based on FIFO, LIFO, TTL, or age [19, 25] of items. By design, however, these strategies see each delivered data item as contributing the same value to the network, and therefore strive to make a similar effort in delivering each item without worrying about ordering. In reality, however, an item's contributed value to different nodes will vary depending on some measure of *relevance*.

In this paper, we claim that all data items are not created equal, and that dissemination systems can benefit from reflecting this reality. One mechanism for a host to determine the relevance of a data item is based on the match between its interests and the item metadata. A node could then decide which items to select for transfer based on the perceived benefit to the recipient or the network as a whole. As the amount of selected data that can be transferred between two nodes is limited by the duration of their contact, the order of the data transferred plays an important role for the performance of the dissemination.

With these observations in mind, we study five data-centric ordering strategies and their ability to deliver relevant data to nodes. By matching user interests against content metadata, recipients assign each data item a relevance score. We then judge a dissemination strategy mainly by its ability to deliver the higher scored data before the lower scored in an efficient manner.

We evaluate each dissemination strategy using the Haggle network architecture [23] running on a trace-driven emulation testbed. Our main results are not specific to Haggle; we simply take advantage of the functionality offered by the platform. We use three different contact traces to study the behaviour of each strategy under varying environments, in terms of connectivity and community structure. Our results show that data transferring based on the relevance leads to a more efficient dissemination of data.

The rest of the paper is organised as follows: Section 2 considers related work, Section 3 presents our formulation of relevance-driven data dissemination. Our experimental details are specified in Section 4 followed by the results in Section 5. Finally, our conclusions are discussed in Section 6.

## 2. RELATED WORK

Nodes cooperating to provide a data dissemination system need to follow mutually beneficial policies in order to achieve their common aims, such as message ferrying, item replication and searching. Opportunistic sharing actions can be divided into four main groups: i) choosing who to *peer* with; ii) *selecting* files to share with peers; iii) *ordering* the transfer of those files; iv) *dropping* files when buffer space runs out. Nodes can follow a wide variety of policies for each of these actions aiming to maximise different properties of the dissemination system.

The problem of choosing who to peer with is fundamental in opportunistic networks, especially in dense or high churn networks. Initiating communication over extremely short-lived connections may prove fruitless, while very long-lived ones need not be prioritised. RAPID is a protocol that considers DTN routing as a resource allocation problem, translating a selected routing metric into per-packet utilities that determine how packets should be selected for replication in the system [2]. Through the learning of node contact patterns they show it is possible to minimise the chosen routing metric. This work allows for specifying the policy for all actions listed at the beginning of this section, though it does not define how other metrics will be affected.

Numerous algorithms have been proposed that select items to forward based on context or social knowledge to enhance the dissemination [5, 6, 10, 24]. However, the goal of such protocols is to compute good "data mules", and are otherwise agnostic to the relevance of the data forwarded.

Forwarding decisions and buffer management in opportunistic networks have been extensively studied. It has been shown that buffer management strategies, e.g., FIFO, LIFO, "most forwarded" or "oldest" can have a large impact on the delay and overhead in a DTN [19]. However, these strategies optimise for high delivery ratio or low delay, and have at most an *incidental* effect on the relevance of the data delivered. Krifa et al. propose a distributed buffer dropping algorithm (the aspect we do not tackle) that uses the theory of encounter-based message dissemination and statistical learning to approximate an optimal approach with global knowledge, improving average delivery rate and delivery delay [14].

A number of data-centric dissemination systems have previously been proposed that use "content channels" as their underlying dissemination mechanism [8, 15, 16]. Data items are classified into channels that users subscribe to and then synchronise when in contact. Content channels provide a coarse definition of relevance, and there is no intentional ordering of data within channels or when synchronising. We use a more specific notion of relevance based upon the metadata of files being shared. Moghadam et al. [20] proposed an interest-aware content distribution protocol with a more fine grained definition of relevance, similar to our relevance score only selecting whether to forward or not, and not how to order data.

Global knowledge of other nodes naturally helps achieve a more efficient dissemination [12], but perfect global knowledge is impractical to achieve. However, strategies that learn as much as possible about other nodes in the system may benefit in the long run, as we show in Section 5.

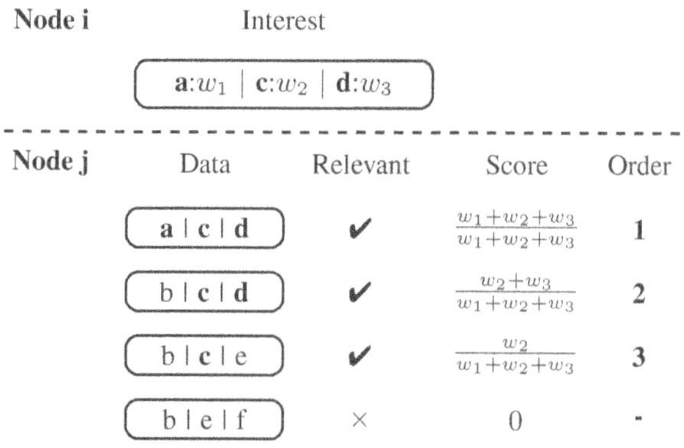

Figure 1: Data ordering according to a node's interests. The recipient $i$ has weighted interests causing the depicted ordering of $j$'s data items.

## 3. RELEVANCE-DRIVEN DATA DISSEMINATION

We now introduce the basic concept of relevance-driven data dissemination, which uses *relevance scores* to decide whether to select items for transfer and which order to send them. The process delivers data items to nodes based on their interests rather than their network identifiers (like traditional networking).

To compute a specific node's relevance score for a data item, we assume that each data item's metadata is in the form of a set of attributes $\mathcal{D}$. The use of metadata is widely adopted, for example in music files' *ID3* tags or image files' *Exif* standard. Likewise, each node possesses a set of interests $\mathcal{I}$ that correspond to data attributes. Interests are forwarded to other nodes through direct contacts or through relaying them via third parties. A node's interests are weighted to emphasise particular interests of a node.

We classify data items as *relevant* to a node if the intersection $\mathcal{R} = \mathcal{D} \cap \mathcal{I}$ is non-empty. The strength of a node's interest in a data item is based on the matching interests $a \in \mathcal{R}$ and their respective weights $w_a$, expressed as a score:

$$score = \frac{\sum_{a \in \mathcal{R}} w_a}{\sum_{a \in \mathcal{I}} w_a} \qquad (1)$$

The relevance score can be used by a dissemination system to determine the ordering in which items to be forwarded should be transferred. Figure 1 shows a transfer schedule from node $j$ to node $i$, based on $i$'s relevance score of the data items carried by $j$. A high relevance score implies that $j$ should transfer the item early in a contact to ensure its successful delivery. Note that this relevance-driven ordering does not determine which encountered nodes are good "data mules", this is the task of a forwarding algorithm.

Although the above example represents a relatively straightforward approach to ordering that locally optimises for the nodes involved in the exchange, other more collaborative approaches, such as exchanging items also relevant to other nodes, may prove more beneficial for the network as a whole and in the long run. Ensuring a data item is maximally relevant to a receiver is an obvious greedy approach for increasing local interest satisfaction. However, in many cases, when the global interest in a data item is strong—although the local interest is weak—it may be preferable to prioritise that item's dissemination over one that maximises the local interest satisfaction. To study these various trade-offs in data dissemination, we define the following strategies for exchanging data items during a node contact:

**ScoreLocal** — Only locally relevant data items are exchanged, ordered by their relevance scores relative to the receiver of a data transfer, with the most relevant data item first.

**ScoreGlobal** — Only globally relevant data items are exchanged, ordered by their aggregate relevance scores relative to all nodes known at the time the data transfer occurs. The purpose of this strategy is to promote the dissemination of globally relevant items over locally relevant ones.

**RandomLocal** — Only locally relevant data items are exchanged, but they are ordered randomly. This is a baseline for evaluating the value of using the relevance score to order data items.

**Random** — Any buffered data items are exchanged in random order, irrespective of relevance to the nodes involved in the data exchange. This strategy provides a baseline for evaluating the value of using the relevance score to both *select* and *order* data items.

**LiFoLocal** — Only locally relevant data items are exchanged, ordered based on buffer time with the "freshest" item first. The purpose of this strategy is to promote fresh data, a property that has been shown to be efficient in buffer management in sparsely connected networks [4].

We shall refer to ScoreLocal, LiFoLocal and RandomLocal as "local strategies" as they select items for transfer based only upon the recipient's interests.

# 4. EXPERIMENTAL METHODOLOGY AND SETUP

In this section we describe the experimental methodology and setup we use to compare dissemination strategies. The strategies are implemented in Haggle [1] and we evaluate them through a set of experiments executed on a trace-driven emulation testbed, where contact traces decide the connectivity between Virtual Machine (VM) nodes [3]. We use three traces with different properties to minimise the bias on our results due to the configuration of a particular trace (Section 4.1), and understand transfer ordering across a range of network densities and churn. We distribute interests and data according to a well-known model of data and interests on the web (Section 4.2). A relevance metric from the area of information retrieval allows us to evaluate the relevance of the data delivered to nodes in our experiments (Section 4.4).

## 4.1 Contact Traces

We use three contact traces, generated from topology/mobility models or real measurements, which each aim to subject our dissemination strategies to different network topologies:

**Clusters in Line** — This is a synthetically generated trace that configures nodes in clusters with the goal of studying dissemination in a segmented network, as shown in Figure 2. The links in-between nodes in each cluster, and links between bridging nodes, are modelled by a two state Markov link model [11], with an expected contact duration of 60 seconds and inter-contact time of 240 seconds. Each of the three clusters is comprised of five nodes.

**HCMM:SO** — This trace is generated using the Home-cell Community based Mobility Model with Social Overlay [9], which aims to accurately model both intra- and inter-community contact patterns. The intra-community pattern is achieved by assigning nodes to home cells (or caves) according to a caveman social graph [22], as illustrated in Figure 3. Nodes then move to another cell $C$

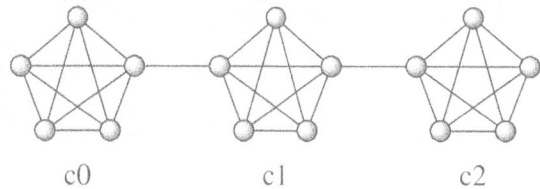

Figure 2: Clusters in Line topology, with three $K_5$ graphs connected by bridge links. Each edge is an intermittently connected Markov link.

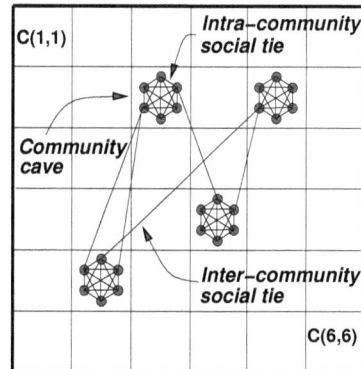

Figure 3: HCMM:SO contact model example: Each node is assigned a home cell and moves to other cells according to its social relations there (source [9]).

with a probability proportional to the number of their "friends" (in the social graph) that have $C$ as their home cell. The inter-community contact pattern is modelled through periods of social activity, in which community *bridge nodes* convene at common locations according to a complementary social graph overlay. This model achieves weak non-homogenous mixing. We use a HCMM configuration with 30 nodes, in 6 communities a 5 nodes distributed over 10x10 cells, and a social overlay with inter-community movement of 2 hours but otherwise the same parameters as in [9]. The nodes have a mean contact duration of 58s, intercontact time of 166s and 38% of the nodes meet each other.

**Real World** — A real-world trace, collected from an experiment with 10 mobile phones, complements our two synthetic traces. The mobile phones were handed out to colleagues in our office corridors, who carried them during their work day. The phones used WiFi to send beacons every five seconds, allowing them to detect each other when in range. The WiFi interfaces used a low transmit power of 2 dBm to reduce battery consumption. The experiment lasted around six hours and people attended their normal work schedules; they sat in their offices, chatted with colleagues, had lunch, and participated in meetings. The nodes have a mean contact duration of 135s, intercontact time of 409s and 82% of the nodes meet each other.

## 4.2 Interest Modeling and Data Distribution

To model a data-centric network, interests and data should be distributed in a way that reflects the non-uniformity of item popularity in real life. Studies on user interest distribution have been shown to exhibit Zipf-like behaviour [7], we therefore model our distribution of interests and content metadata accordingly. To this end, we form node interests by assigning each node a set of five

55

| Parameter | Symbol | Default Size |
|---|---|---|
| Attribute pool | $\mathcal{A}$ | 100 |
| Data attributes | $\mathcal{D}$ | 5 (Zipf, $\alpha = 0.5$) |
| Interest attributes | $\mathcal{I}$ | 5 (Zipf, $\alpha = 0.5$) |
| Data items per node | | 200 |

Table 1: Parameters used when generating (meta)data and interests.

| Zipf $\alpha$ | 0.25 | | | 0.5 | | | 0.75 | | |
|---|---|---|---|---|---|---|---|---|---|
| dst\src | c0 | c1 | c2 | c0 | c1 | c2 | c0 | c1 | c2 |
| c0 | 100 | 13 | 0 | 100 | 19 | 1 | 100 | 30 | 17 |
| c1 | 15 | 100 | 17 | 14 | 100 | 24 | 23 | 100 | 64 |
| c2 | 0 | 16 | 100 | 0 | 20 | 100 | 13 | 65 | 100 |

Table 2: Fraction of inter-cluster data delivered for different Zipf $\alpha$ in the *Clusters in Line* topology, using any local strategy.

attributes $\mathcal{I}$ from a global attribute pool $\mathcal{A}$ of cardinality 100. Each user's interest attributes are drawn from $\mathcal{A}$, which is a Zipfian distribution with skewness parameter $\alpha$. A higher skewness leads to a higher overlap of the interest between the nodes, while $\alpha = 0$ corresponds to a uniform distribution of the interests. This method allows to control the strength of the interest segregation among nodes. Each attribute $a \in \mathcal{I}$ is also given a weight $w_a$, proportional to the likelihood of picking $a$ from $\mathcal{A}$, and normalised such that the sum of a node's interest weights equals 100.

We assign each data item a set of five attributes $\mathcal{D}$ from the attribute pool $\mathcal{A}$ with the same probability distribution as above. The node interests and item attributes use the same distribution, as libraries generally reflect the tastes of the owners. We chose to inject all data items at the beginning of the experiments to enable complete analysis of a single wave of item distribution and so that each data item has equal conditions (i.e., node contacts) to disseminate.

We explore different values of $\alpha$ for the attribute pool $\mathcal{A}$, unless otherwise specified we use $\alpha = 0.5$ as a default. This reflects a balanced attribute overlap, avoiding a situation where all nodes have interest in every data item. Instead, with our configuration, a node has an interest in $\sim 22\%$ of all data items. The experiment parameters are summarised in Table 1.

### 4.3 Data Transfer

To focus on the transfer ordering effect when forwarding, rather than whether an item is selected for forwarding, we artificially limit the number of transfers per node contact. This to emphasise the impact of limited contact duration, and to abstract away from Bluetooth or WiFi bandwidth offered by mobile phones, data size, and the distribution of contact duration in the given contact traces.

The results presented in the following section are derived with a limit of 10 data objects transferred per node contact. Different configurations with up to 100 data objects per node contact showed comparable relations between the different strategies, we therefore concentrate the analysis on one configuration. Furthermore, a low number of transfers reflects resource conservative nodes, a likely situation in real life.

### 4.4 Evaluation Metrics with Relevance Focus

To measure the effectiveness of relevance-driven dissemination we use the normalised Discounted Cumulative Gain (nDCG) [13], an established metric in the information retrieval community. The nDCG assigns a given result set a quality valuation between 0 and 1, and assesses the ordering of data items in a set compared to an ideal order. In our case, the ideal ordering is obtained for a node by taking the top $T$ sorted (according to relevance) data items out of all possible items, while the experienced ordering is the top $T$ sorted sequence of items that have been delivered to the node in the experiment. The nDCG metric thus gives a measure of how successful a strategy is at delivering data of high relevance. The nDCG is computed as follows:

$$nDCG^n[T] = n_T(s_1 + \sum_{i=2}^{T} s_i/log_b(i)) \qquad (2)$$

where $s_i$ is the score of the $i^{th}$ relevant received data item (i.e., its local order is $i$) and $1/n_T$ is the $nDCG^{n-ref}[T]$ for the relevant data available in the system used for normalisation. This measure will only consider how well the system provides a node's $T$ most desired data items, items below rank $T$ will not contribute utility to the node. The network-wide nDCG is calculated as the mean of the per-node $nDCG^n$, and gives a measure of a strategy's performance for the network as a whole. It can be seen that all nodes are weighted equally to this systemic measure, that is, nodes with low interest in available items contribute equally to the network-wide score.

The nDCG does not give a complete picture of the performance of a strategy, revealing nothing about the cost to achieve a certain nDCG or whether less relevant data is unnecessarily delayed in order to achieve a high nDCG. We therefore complement our analysis with traditional metrics, such as *delivery ratio*, *delay* and *overhead*. However, these metrics are well understood and require no further introduction.

## 5. RESULTS

In this section we look at the ability of dissemination strategies to deliver relevant items, the delay items suffer and how these metrics evolve through the experiment. This allows us to understand the learning process of globally oriented strategies, which require time to learn about the interests of other nodes. The cost of dissemination (i.e., overhead) also reveals important trade-offs between performance and transfer efficiency.

### 5.1 Delivery Ratio

Each node is interested in a subset of the data distributed across the nodes in the network. A single node's delivery ratio is the fraction of relevant items that have been received by the node. The network-wide delivery ratio is the mean node-specific delivery ratio. It gives an indication of how successful the network has been at delivering data items of relevance to all nodes throughout the network.

*Dissemination Across Segmented Networks.*

The distribution of data items and interests, the contact patterns, and the length of the experiment all constrain the delivery process. In particular, interest and data distributions affect strategies that only make local considerations. These strategies suffer when networks are segmented, because nodes isolated from each other must rely on other nodes to transfer the data they desire from other parts of the network. With local strategies, such transfers only occur when the relevant data items are also of interest to the recipient.

The *Clusters in Line* topology helps us to study the effect of segmented networks on local strategies. The delivery ratio across clusters is bound by the interests of the bridge nodes; only data of in-

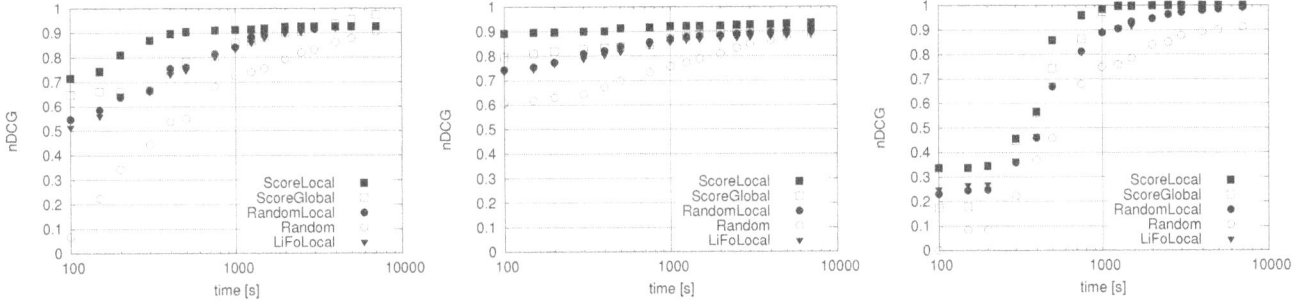

Figure 4: Normalized Discounted Cumulative Gain nDCG evaluated over the 10 most relevant received data over time of the experiment for (a) Clusters in Line (b) HCMM:SO (c) Real World

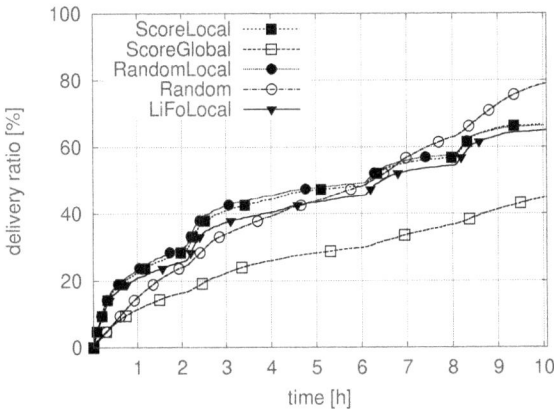

Figure 5: HCMM:SO: Delivery ratio progress over time, $\alpha = 0.5$.

terest for the bridge node in cluster c0 will be disseminated from cluster c1 to cluster c0, for example. By changing the $\alpha$ parameter of the Zipf distribution, we can study how the node interests affect the dissemination. Table 2 shows the fraction of items delivered from one cluster to another. About 15% of relevant data items are able to traverse to adjacent clusters when $\alpha = 0.25$. No items traverse between the two non-adjacent clusters (c0 and c2). Increasing $\alpha$ to 0.5 allows one percent of the items to traverse from c0 to c2, also slightly improving adjacent traversal. Raising $\alpha$ to 0.75 improves the non-adjacent dissemination further.

As local strategies only transfer locally relevant data, every transferred item contributes to the delivery ratio independent of the ordering strategy. While the strategies eventually achieve the same delivery ratio, their progress is different as the availability of relevant data items becomes scarce. The next section discusses this aspect.

*Delivery Ratio Progress.*

The rate at which relevant items are delivered gives an indication of the strategies' ability to prioritise the network-wide delivery ratio. HCMM:SO periodically exposes nodes to new neighbours, allowing us to examine the effect of the local saturation of neighbours' items. We can see this in Figure 5, where the progress of delivery ratio is given for all strategies. Local strategies perform well initially but start to experience limited availability of items left to share, reducing progress until new nodes are encountered (as seen at hours 2, 6, and 8). This pattern is repeated throughout the

experiment. Random and ScoreGlobal strategies select more data items, and therefore suffer less from this local saturation effect.

Successful delivery of new items will be constrained when all contacts are used to distribute the same data, the network needs variety to ensure all interests are satisfied. This can be seen in the slow growth of ScoreGlobal, which aims to have all nodes swapping the same "best" items, starving delivery of less globally relevant items. For this reason, Random begins to outperform the others after 7 hours, as it enforces variety in dissemination among nodes.

Variety in data among the nodes is in particularly important when local saturation affects the ability to progress in data delivery. LiFoLocal performs worst of the Local strategies in terms of delivery ratio because promoting fresh data does not improve variety in data among the nodes the same way as random or score-based ordering do.

## 5.2 Received Item Relevance

The delivery ratio by itself does not give any information about the ability of a strategy to disseminate the *most* relevant data. We measure the relevance of the data received by nodes compared to the most relevant data available in the system using the nDCG metric. Figure 4 shows the network-wide nDCG progress evaluated over the 10 most relevant received data items for the three topologies. At the beginning of all experiments, ScoreLocal consistently delivers the most relevant data, as the strategy was designed to exchange the most relevant data first. In the *Clusters in Line* topology we see the ScoreGlobal strategy eventually exceed the performance of ScoreLocal due to network segmentation from which local strategies suffer. It is also interesting to note that ScoreGlobal develops its knowledge about other nodes over time, at the first node contact it is equivalent to ScoreLocal, as it does not have previous knowledge about other nodes. Thus the strategy initially achieves a higher nDCG than random or freshness based strategies but loses this behaviour over time as it meets more nodes and develops a global view. The HCMM:SO results begin high due to the large amount of early contacts.

In terms of relevance, using data freshness as an ordering criteria does not give an advantage over random ordering, so the promotion of newly received items does not have the same beneficial effect as it does in buffer management [4]. Moreover, its tendency to distribute the same data to many nodes can lead to worse results than random ordering due to excessive redundancy.

*Interest Segregation.*

The effect of varying $\alpha$ for the *Clusters in Line* topology is depicted in Figure 6. For high $\alpha$, most nodes and items will share

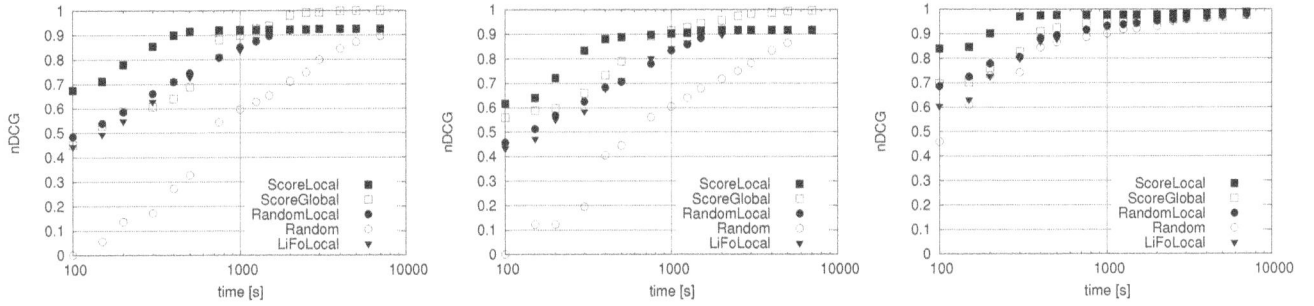

Figure 6: Normalized Discounted Cumulative Gain nDCG on *Clusters in Line* topology evaluated over the 10 most relevant received data over time of the experiment for varying values of $\alpha$ (a) $\alpha = 0$ (b) $\alpha = 0.25$ (c) $\alpha = 1$

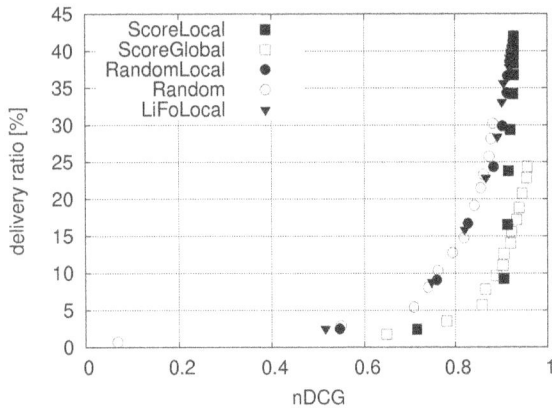

Figure 7: Clusters in Line: Normalized Discounted Cumulative Gain nDCG versus delivery ratio.

| Zipf $\alpha$ | 0 | 0.25 | 0.5 | 0.75 | 1 |
|---|---|---|---|---|---|
| *Local | 1.00 | 1.00 | 1.00 | 1.00 | 1.00 |
| ScoreGlobal | 3.09 | 2.98 | 2.41 | 1.51 | 1.08 |
| Random | 3.77 | 3.63 | 2.97 | 1.86 | 1.24 |

Table 3: Clusters in Line: Overhead in terms of number of transferred data items per received relevant data item.

the more likely attributes. When $\alpha$ is low, attributes are evenly distributed and so it will be unlikely that two adjacent nodes find the same item relevant. This lack of shared relevance limits items' ability to spread with a local strategy. We can see the effect of this in Figure 6, where the local strategies do not reach 100%. This is due to the segregation of nodes with a matching interest in items. Whereas non-local strategies can overcome these limitations, as transferred items do not need to interest the recipient.

With sparse topologies the path length to other nodes will be longer than in dense topologies. Longer paths are less likely for all adjacent nodes to share interest in data items. Thus shorter path lengths will not suffer from as much segregation.

As ScoreGlobal develops a more informed view of other nodes' interests, it will gain a more uniform view of interest distribution. A lack of strong views on what should be shared avoids segregation.

Informed ordering of data items is of particular concern when nodes have particularly segregated interests (low $\alpha$). Whereas with highly skewed interests (high $\alpha$) the Random strategy does not suffer from its simplicity.

*Relevance versus Delivery Ratio.*

Comparing relevance against the delivery ratio gives an indication about the efficiency of the strategies. While, achieving a high nDCG with low delivery ratio is likely to involve transfer of irrelevant data, it shows that the data delivered actually is the most relevant one.

In Figure 7, we plot the nDCG against the delivery ratio. Each data point represents periodic instances in time as both delivery ratio and nDCG monotonically increase over the experiment. More separation between points thus shows better progress towards delivery and relevance.

Random and freshness-based ordering show the same behaviour, while the Random strategy has slowest progress. Relevance-based ordering naturally achieves higher nDCG for the same delivery ratio than other orderings, with the global selection strategy ScoreGlobal overcoming the high relevance/low delivery property slower. This indicates the increased cost of achieving its higher relevance score at the end of the experiment. ScoreLocal on the other hand cannot take advantage of the increasing delivery ratio to improve its nDCG score due to the previously discussed interest segregation.

## 5.3 Transfer Efficiency

We use the overhead in the number of transferred data items per received relevant data item as measure of efficiency. Table 3 summarises the overhead for different data an interest distributions for the *Clusters in Line* topology. While the overlap in data and interest distribution affects the overhead needed to disseminate the data, we found that the results are similar for the three topologies. Local strategies achieve optimal overhead of 1.0 as every transferred data item is relevant to its recipient.

Both ScoreGlobal and Random try to benefit the whole system and therefore use contact opportunities to transfer data that is not necessarily relevant to its recipient. As the interests of the recipient are included in the aggregated interests used to select data, the chances of transferring relevant data items are higher for Score-Global than using the Random strategy. Looking at the data and interest distribution, we observe that increased skewness leads to a reduction in overhead. This is because more nodes are interested in the same data item, increasing chances that a recipient happens to be interested in the item. The overhead of the ScoreGlobal strategy almost reaches optimality with a skewness of $\alpha = 1$, reflecting that the individual node interests are highly overlapping given that data and interest distribution.

# 6. CONCLUSIONS

This paper proposed that the ordering of data items in transfer limited data-centric opportunistic networks has a large impact on the relevance of received data. We experimentally compared five strategies that select and order data to be transferred and evaluated their ability to deliver the most relevant data, measured using the normalised Discounted Cumulative Gain (nDCG). The trace-driven experiments on an emulation testbed included contact traces with a range of densities and mixing.

The selection and ordering of transferred items significantly affects the achieved delivery ratio and its rate of increase, respectively. If nodes desire highly relevant data, careful ordering of limited transfers can satisfy these desires quicker than order agnostic approaches. With local strategies, if it is unlikely that neighbours are interested in the same content, segregation can occur where nodes are unable to disseminate items through the network. Interestingly, considering the relevance of data provides a way to overcome this segregation.

In the case of densely connected networks, just local decisions can achieve system properties. However, in particularly sparse networks we need to explicitly consider non-local information to successfully promote system concerns. We saw how selecting items to transfer using global information eventually outperforms local approaches. However, this does not mean that changing strategies in the experiment achieves the best of both approaches, as the early transfers dictate later performance.

We investigated limited transfer capabilities in data-centric opportunistic dissemination systems. We propose that the ordering of item transfers is a behaviour that, while rarely explicitly considered, has a large impact on the timely delivery of the most relevant data. A fully developed dissemination system should still consider more aspects than just relevance. Variable item size, contact duration and buffer size would complicate ordering decisions. It may then be advantageous to consider how best limited capacity and storage should be used [17].

For future work, we plan to consider injecting data into nodes continuously, the correlation between node topology and interest structures, and an analysis of item diversity and fairness aspects.

## Acknowledgements

This research was funded by the ResumeNet project under the EU grant FP7-224619. We thank Theus Hossmann for making the HCMM:SO code and Figure 3 available to us.

# 7. REFERENCES

[1] Haggle code project page, http://haggle.googlecode.com.

[2] BALASUBRAMANIAN, A., LEVINE, B., AND VENKATARAMANI, A. DTN Routing as a Resource Allocation Problem. In *ACM SIGCOMM Computer Communication Review* (2007), vol. 37, ACM.

[3] BJUREFORS, F., GUNNINGBERG, P., AND ROHNER, C. Haggle Testbed: a Testbed for Opportunistic Networks. In *7th Swedish National Computer Networking Workshop* (2011).

[4] BJUREFORS, F., GUNNINGBERG, P., ROHNER, C., AND TAVAKOLI, S. Congestion Avoidance in a Data-Centric Opportunistic Network. In *ACM SIGCOMM ICN Workshop* (2011).

[5] BOLDRINI, C., CONTI, M., IACOPINI, I., AND PASSARELLA, A. HiBOp: a History Based Routing Protocol for Opportunistic Networks. In *IEEE WoWMoM* (2007).

[6] BOLDRINI, C., CONTI, M., AND PASSARELLA, A. ContentPlace: Social-Aware Data Dissemination in Opportunistic Networks. In *ACM MSWiM* (2008).

[7] BRESLAU, L., CUE, P., CAO, P., FAN, L., PHILLIPS, G., AND SHENKER, S. Web Caching and Zipf-like Distributions: Evidence and Implications. In *INFOCOM* (1999).

[8] HELGASON, O. R., YAVUZ, E. A., KOUYOUMDJIEVA, S. T., PAJEVIC, L., AND KARLSSON, G. A Mobile Peer-to-Peer System for Opportunistic Content-Centric Networking. In *MobiHeld* (August 2010).

[9] HOSSMANN, T., SPYROPOULOS, T., AND LEGENDRE, F. Putting Contacts into Context: Mobility Modeling beyond Inter-Contact Times. In *MobiHoc* (Paris, France, May 2011).

[10] HUI, P., CROWCROFT, J., AND YONEKI, E. Bubble Rap: Social-based Forwarding in Delay Tolerant Networks. In *MobiHoc* (2008).

[11] HWANG, S., AND KIM, D. Markov model of link connectivity in mobile ad hoc networks. *Telecommunication Systems 34* (2007).

[12] JAIN, S., FALL, K., AND PATRA, R. Routing in a Delay Tolerant Network. *ACM SIGCOMM CCR 34*, 4 (2004).

[13] JÄRVELIN, K., AND KEKÄLÄINEN, J. IR evaluation methods for retrieving highly relevant documents. In *ACM SIGIR* (2000).

[14] KRIFA, A., BARAKA, C., AND SPYROPOULOS, T. Optimal Buffer Management Policies for Delay Tolerant Networks. In *SECON* (2008), IEEE.

[15] KRIFA, A., BARAKAT, C., AND SPYROPOULOS, T. MobiTrade: Trading Content in Distruption Tolerant Networks. In *CHANTS* (September 2011).

[16] LENDERS, V., KARLSSON, G., AND MAY, M. Wireless Ad Hoc Podcasting. In *IEEE SECON* (June 2007).

[17] LI, Q., ZHU, S., AND CAO, G. Routing in Socially Selfish Delay Tolerant Networks. In *INFOCOM* (2010), IEEE.

[18] LINDGREN, A., DORIA, A., AND SCHELÉN, O. Probabilistic Routing in Intermittently Connected Networks. *SIGMOBILE MCCR* (2003).

[19] LINDGREN, A., AND PHANSE, K. Evaluation of Queueing Policies and Forwarding Strategies for Routing in Intermittently Connected Networks. In *Comsware* (2006).

[20] MOGHADAM, A., AND SCHULZRINNE, H. Interest-Aware Content Distribution Protocol for Mobile Disruption-Tolerant Networks. In *WoWMoM* (2009).

[21] MTIBAA, A., MAY, M., DIOT, C., AND AMMAR, M. PeopleRank: Social Opportunistic Forwarding. In *IEEE INFOCOM* (2010).

[22] NEWMAN, M. E. J. The Structure and Function of Complex Networks. *SIAM REVIEW 45* (2003).

[23] NORDSTRÖM, E., GUNNINGBERG, P., AND ROHNER, C. A Search-based Network Architecture for Mobile Devices. Tech. Rep. 2009-003, Department of Information Technology, Uppsala University, Jan. 2009.

[24] SOLLAZZO, G., MUSOLESI, M., AND MASCOLO, C. TACO-DTN: A Time-Aware Content-based Dissemination System for Delay Tolerant Networks. In *MobiOpp* (2007).

[25] SPYROPOULOS, T., PSOUNIS, K., AND RAGHAVENDRA, C. S. Spray and Wait: An Efficient Routing Scheme for Intermittently Connected Mobile Networks. In *ACM SIGCOMM WDTN Workshop* (2005).

[26] VAHDAT, A., AND BECKER, D. Epidemic Routing for Partially-Connected Ad Hoc Networks. Tech. rep., 2000.

# A Disruption-tolerant Transmission Protocol for Practical Mobile Data Offloading

Younghwan Go, YoungGyoun Moon, Giyoung Nam, and KyoungSoo Park

Department of Electrical Engineering, KAIST
Daejeon, South Korea
{yhwan, ygmoon, giyoung}@ndsl.kaist.edu, kyoungsoo@ee.kaist.ac.kr

## ABSTRACT

The explosive popularity of smartphones and mobile devices drives massive growth in the wide-area mobile data communication. Unfortunately, the current or near-future 3G/4G networks are deemed insufficient to meet the increasing data transfer demand. While opportunistic offloading of mobile data through Wi-Fi is an attractive option, the existing transport layer would experience frequent disconnections due to mobility, making it hard to support seamlessly reliable data delivery. As a result, many mobile applications either depend on ad-hoc downloading resumption mechanisms or redundantly re-transfer the same content when disruptions happen.

In this paper, we present DTP, a disruption-tolerant, reliable transport layer protocol that masks the failures of the preferred network. Unlike previous disruption/delay-tolerant protocols, DTP provides the same semantics as TCP on an IP packet level when the mobile device is connected to a network while providing the illusion of continued connection even if the underlying physical network becomes unavailable. This would help the mobile application developers to focus on the application core rather than addressing the frequent network disruptions. It would also greatly reduce the phone network costs both to ISPs and end users. Our current implementation in UDP shows a comparable performance to that of TCP in network, and it greatly reduces the delay and power consumption when the mobile devices frequently switch from one network to another.

## Categories and Subject Descriptors

C.2.1 [**Network Architecture and Design**]: Store and forward networks; C.2.1 [**Network Architecture and Design**]: Wireless communication; C.2.2 [**Network Protocols**]: Applications

## General Terms

Design, Performance

## Keywords

Delay Tolerant Network, Wi-Fi Offloading, Mobility

## 1. INTRODUCTION

Recent advancement in cell-phone networks and smartphones has brought massive growth in the mobile data communication. The number of mobile network users is expected to surpass that of the wired Internet within the next four years [1] and the global traffic volume is predicted to consume 6.3 Exabytes per month in 2015, a 26-fold increase from that of 2010 [2]. However, the existing 3G or Long Term Evolution (LTE) networks in the near future are unlikely to provide as much bandwidth as in the wired Internet, and the capacity shortage is becoming a serious barrier to advancing the mobile data communication.

There have been a number of works that address the capacity overloading problem. One end of the efforts is to increase the physical capacity by reducing the cell size or by intelligent multiplexing of the shared radio medium [3,4]. However, these approaches have fundamental limitations when the aggregate network demands exceed the physical capacity. The other end of the line focuses on adopting the hybrid usage of 3G and much higher-bandwidth networks such as the wired Internet through Wi-Fi. The idea is to offload the 3G mobile data transfer via Wi-Fi opportunistically while using the 3G networks as a backup medium to meet the transfer deadline [5–8]. This opportunistic Wi-Fi offloading is an attractive option especially in urban areas with high Wi-Fi availability, with the potential of reducing the 3G data bandwidth consumption via delay-tolerant networking (DTN) [9]. We believe that many non-interactive data-intensive applications such as podcast [10,11], TV episode or movie downloading [12], or personal storage synchronization [13] could benefit from it.

In this paper, we promote delay-tolerant, opportunistic Wi-Fi offloading of 3G mobile data from a practical point of view. Our goal is to support Wi-Fi offloading with little or no change to the current applications or underlying networks. We observe that the 3G or Wi-Fi networks show stable behavior most time while the mobile devices are connected, but one needs to handle network disruptions preferably in a transparent manner when the mobile devices switch from one network to another. According to recent measurements, 87% of the entire smartphone usage occurs while the users are on the move, implying frequent switches between multiple networks [14].

One approach is to have the applications handle network disruptions by themselves. In fact, some applications already support download resumption when they change their network attachments. However, this necessitates an ad-hoc implementation of download resumption in each application (e.g., HTTP byte-range queries, CGI parameter passing, and so on), which cannot be easily reused by other applications. For dynamically-generated contents, applications may not be able to determine where to resume downloading or end up with re-downloading the whole content on a new connection if the IP address of the device changes. Another approach, which we favor in this work, is to transparently handle the network disruptions in the transport layer. Since the majority of mobile applications use TCP, if we make TCP disruption-tolerant, many non-interactive applications could benefit from transparent Wi-Fi offloading with minimal change. This would also ease the burden on the application developers so that they focus on the core program logic rather than handling network failures or disruptions due to device mobility.

We present the design and implementation of DTP, a disruption-tolerant transport layer protocol that transparently masks network failures from the application layer. On a high level, DTP works similarly to TCP when the mobile device is attached to a network but it provides the illusion of continued connection to the applications even when the underlying network is unavailable. This way, DTP allows the mobile applications to exploit Wi-Fi offloading without requiring them being DTN-aware. Unlike previous DTN protocols [15–18], DTP supports reliable data delivery on a packet level and it does not require any special support from the network infrastructure.

The key technical challenge in DTP is how we manage the connection when the physical network switches between on and off. Instead of binding the connection on the four connection tuples (source and destination IP addresses and port numbers), DTP binds the connection to a *flow ID* that is agreed at the initial connection setup time and does not change during the connection lifetime. When a mobile host moves to another network, it can resume the connection with a new IP address and a port number by cryptographically attesting that it owns the flow ID of the connection. The DTP connection closes either when both parties explicitly tear it down or when the *keep-alive duration* of the connection expires. The keep-alive duration is the estimated connection lifetime set at the connection setup time that can be updated during the course of the connection.

While DTP hides the network disruptions transparently from the application layer, it presents a few security problems. Malicious hosts may attempt to hijack a connection by resuming an interrupted one or create lots of fake states on the server side. To prevent connection hijacking, DTP exchanges a secret key at connection setup and authenticates the other end by a simple challenge-and-response protocol before resuming. To mitigate the state explosion attacks, DTP keeps a minimal state per flow (less than 200 bytes per flow), reducing the memory burden on the server.

We build the prototype of DTP as a UDP-based API library where each function has a one-to-one correspondence to a TCP socket function. Our initial evaluation shows that its performance is comparable to that of TCP on wired or Wi-Fi networks while it shows 47% and 123% better performance for moderate and large file sizes in a typical delay-tolerant setting.

## 2. BACKGROUND

Many smartphones and tablet PCs these days have both 3G and and Wi-Fi interfaces. The availability of 3G is typically more ubiquitous than that of Wi-Fi [5], but it is more expensive due to smaller capacity. While the next generation cell-phone networks such as LTE are being deployed, they are unlikely to catch up with the fast-growing mobile data demand in the future.

To mitigate the 3G capacity overloading problem, many data-intensive mobile applications configure Wi-Fi as the preferred interface by default, and explicitly ask for the permission from the user when it needs to switch to 3G. Recent studies show that one can benefit further from Wi-Fi offloading if we allow some delay

| Category | 3G | Wi-Fi |
|---|---|---|
| Availability | 100% | 45%(Vehicle) / 53%(Walk) |
| Latency | 130 ms | 80 ms |
| Bandwidth | 1 - 2 Mbps | 2.6 - 5 Mbps |

**Table 1: Wi-Fi Availability Test in Visiting a Large City**

for data transfer [5, 6]. We can offload 10% to 30% of 3G data to Wi-Fi if we disallow any interruptions in data transfer, but the offloading ratio can go up to 75% if we allow 30 minutes of delay and to 88% for 6 hours of delay [6]. While delays in real-time contents would lead to poor user experience, we find that many non-interactive applications (e.g., large-file downloading) could benefit from delayed Wi-Fi offloading and reduce monthly 3G data bills. Moreover, some delays could present opportunities for intelligent load balancing in the network itself by shifting the bandwidth usage to a less congested timeframe.

Our work bridges the previous studies with practical offloading support from the transport layer. To facilitate the delayed transfer with as little burden on the application developers as possible, we propose using a TCP-like transport layer that is resilient to network disruptions or failures. In this section, we first check the feasibility of Wi-Fi offloading and the current Wi-Fi offloading practice with popular mobile applications.

### 2.1 Opportunities for Wi-Fi Offloading

We gauge the viability of Wi-Fi data offloading by measuring the availability, connection time, inter-arrival time, bandwidth and latency in a large city in South Korea. First, we have three people measure the 3G/Wi-Fi availability by visiting popular places in Seoul for 4 days. Second, we draw the similar statistics from previous Wi-Fi/3G usage traces of 97 iPhone users for 18 days [6]. Our measurements are by no means representative, but we believe that they show some sense of feasibility of Wi-Fi-based 3G data offloading in an urban setting.

#### 2.1.1 Wi-Fi Availability in Visiting a Large City

To see how widely Wi-Fi is available in a casual visit to a large city, we measure the Wi-Fi availability, bandwidth and latency in a few popular places in Seoul. We pick four popular places (Gang-nam, Myongdong, Insa-dong (all outdoors), and Co-Ex (indoors) in Seoul) where many people visit, and move between them by public transportation. During the 4-day visit, we gathered the data for 27 hours (including 6.7 hours on the subway, and 4.4 hours on a bus). We find that Wi-Fi is available for Internet access for about 45% of the time either on the subway or on a bus (73% on the subway, 5% on a bus), and for 53% of the time while walking around the popular places. Most of these Wi-Fi hotspots are provided by a few ISPs in South Korea. For example, one of the ISPs claims that it has over 87K hotspots nationwide [19].

We also measure the bandwidth and latency from these locations to a server in our lab (about 200 km away from the locations) by implementing a simple Android application. Whenever an Android client meets and connects to a new Wi-Fi AP, it calculates the latency between the client and the server by recording the minimum time it takes to receive a response by a ping request. Next, we measure the bandwidth by transmitting a large file to the server and calculating the total transfer time. Table 1 shows that the Wi-Fi bandwidths are between 2.6 to 5 Mbps on average while those of 3G are from 1 to 2 Mbps. than The average latency of Wi-Fi and 3G are about 80 ms and 130 ms. The Wi-Fi latency looks a bit high presumably because it is used by many people in the areas. Overall,

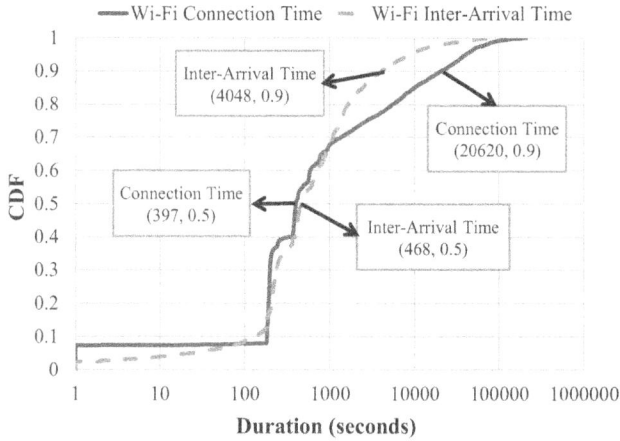

**Figure 1: CDF for Wi-Fi Connection and Inter-Arrival Times**

| Application | Category | Resumption method |
|---|---|---|
| Dropbox | Online storage | Not Supported |
| MapDroyd | Offline map access | Not Supported |
| Winamp | Podcast manager | Not Supported |
| Android Market | App. downloading | HTTP Range Request |
| Beyondpod | Podcast manager | HTTP Range Request |
| Google Listen | Podcast manager | HTTP Range Request |
| TubeMate | YouTube video | CGI Parameter Passing |

**Table 2: Download Resumption in Popular Mobile Applications**

our results show that Wi-Fi has larger available bandwidths even in busy places in a large city.

### 2.1.2 *Wi-Fi Availability in Daily Lives*

We analyze the Wi-Fi availability in daily lives in a large city. We use the traces of 97 iPhone users who periodically measure the Wi-Fi availability in their daily lives in Seoul for 18 days [6]. We measure the connection and inter-arrival time of Wi-Fi by checking the network status of the client every three minutes. Figure 1 shows the distributions of Wi-Fi connections and inter-arrival times in CDF. The graph shows that about the half of the connection times are less than 6.6 minutes while 10% of them show more than 5.7 hours. The long connection times are mostly for staying at home or at work while small ones indicate transient Wi-Fi availability on the move. Also, about the half of the inter-arrival times are less than 7.8 minutes and 10% of them are over 1.1 hour. This result implies that while there are many short Wi-Fi connections, the inter-arrival times are also small, expecting to meet a new Wi-Fi spot soon.

We find that there are good opportunities for Wi-Fi offloading but in order to maximize the benefit, we need to exploit frequent network disruptions and re-connections to our advantage. In our measurements, we observe that without proper upload or download resumption, the users cannot send or receive a file larger than 120 MB for the half of the Wi-Fi connections and would simply waste the Wi-Fi bandwidth and battery power.

## 2.2 Mobile Applications in Disruptions

To examine how current mobile applications respond to network disruptions, we analyze the behavior of seven popular Android applications that are downloaded more than 100,000 times from Android market [20] as shown in Table 2. Dropbox [13] provides online storage synchronization, MapDroyd [21] is used to download and store world-wide maps for offline access. Beyondpod [22], Google Listen [23], and Winamp [24] are audio players that can be linked with podcasting services. TubeMate [12] is a video downloader for YouTube.

When mobile devices experience network disconnections during data transmission, Dropbox, MapDroyd, and Winamp stop with a network failure message, and the downloads do not resume even when the network becomes available again. We find that Android Market, Beyondpod, and Google Listen support download resumption by HTTP byte-range queries while TubeMate uses CGI parameter passing. Even though these applications support download

resumption at network disruptions, there is no common library or rules that can be reused for other applications. Also, it is unclear how one supports the streaming contents where the data is generated on the fly.

Even when applications do not implement download resumption, TCP will resume the connection if a disruption clears up within the time for the maximum number of retransmissions of the same segment assuming the IP address does not change. The total time for retransmission before disconnection is recommended to be larger than 100 seconds [25]. We find that the number of maximum retransmissions is 15 on Linux (kernel version 2.6.40), which corresponds to about 17 minutes. This implies that TCP on Linux cannot handle network disruptions longer than 17 minutes even if the IP address does not change. We design DTP to overcome frequent disruptions and IP address change in mobile environments.

## 3. DESIGN

This section describes the design of DTP, a disruption-tolerant, reliable transport layer protocol. An example is shown in Figure 2. We present the basic protocol, security features, and failure recovery mechanisms at network disruptions.

## 3.1 Disruption-tolerant Connection

TCP binds the IP addresses and port numbers (or the four tuples) of the two communicating ends on its connection. If any one of them changes, the connection needs to be re-established to resume the data transfer. This implies that the IP address of a TCP connection is used to identify the host location of the network as well as the host itself. This duality of the IP address, however, creates a problem in mobile environments where the host location changes frequently due to host mobility.

In order to maintain a connection despite network disruptions or IP address change, DTP binds a connection to a special identifier called "flow ID". The flow ID is determined at initial connection setup time, and it uniquely identifies the connection on both hosts. The DTP connection persists until either it is explicitly torn down by both ends or the keep-alive duration expires. The keep-alive duration is an estimated connection lifetime set by the application, during which the connection stays on even without the physical network availability. We discuss the details in section 3.2.

If the IP address of a mobile host changes, DTP sends a packet with a new IP address to the other end to initiate the authentication process that proves the ownership of the flow ID. If the other end identifies the connection for the flow ID, both parties can resume data transfer from where it left off. Disruption-tolerant connections bring several advantages. First, it allows the application developers not to worry about frequent network disruptions in the Wi-Fi offloading scenarios. They can assume that the connection is always on until it is done with data transfer. Second, it enables the same

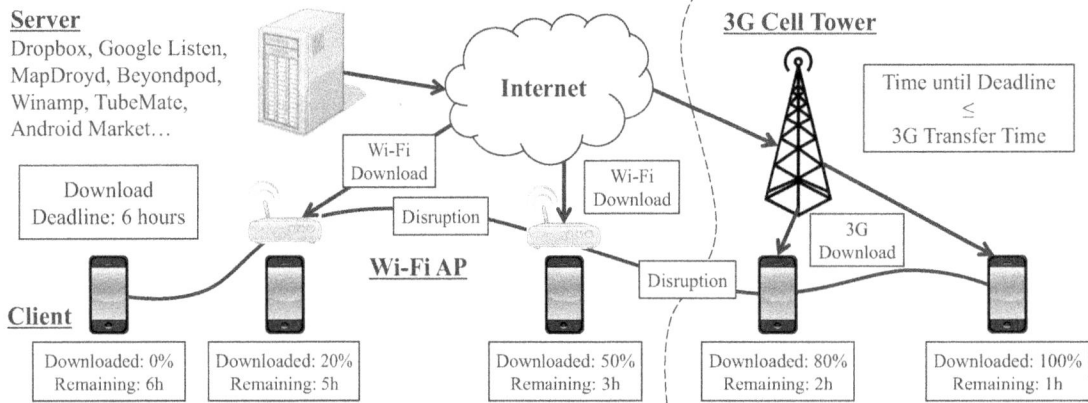

**Figure 2: An Example of a Delay/Disruption-Tolerant Network with Wi-Fi Offloading. The client begins downloading from the server with a 6-hour deadline. The file is downloaded through Wi-Fi whenever the client comes in contact with an AP. If the remaining deadline time is less than or equal to the expected 3G transfer time, the client switches from Wi-Fi and downloads through 3G.**

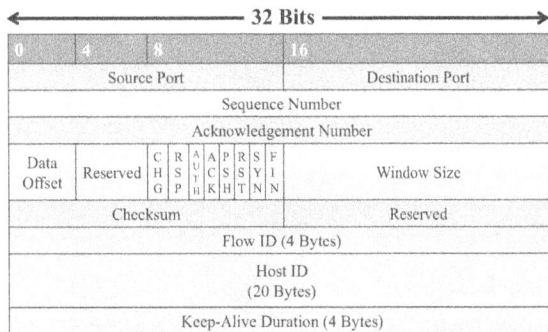

**Figure 3: DTP Protocol Header**

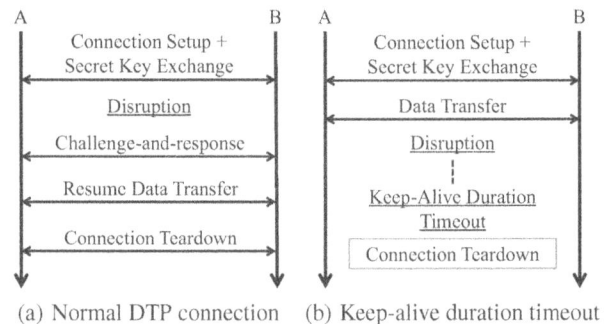

(a) Normal DTP connection    (b) Keep-alive duration timeout

**Figure 4: DTP Communication Timeline**

connection to switch between Wi-Fi and 3G without notice from the application. It allows seamless offloading of even interactive or real-time data without re-establishing the connection or download resumption by the application. One downside of DTP, however, is in the burden to maintain the connection state even when the host is unavailable. Our prototype DTP implementation requires only 176-byte state information per connection, so even for one million concurrent connections, the system would need less than 200 MB for the states.

## 3.2 DTP Protocol Header

The DTP protocol header is shown in Figure 3. We borrow most required fields from the TCP header. The CHG, RSP, and AUTH bit flags are used for secret key validation and will be explained in section 3.3.2. In the option fields, we define the flow ID as the last four bytes of the SHA-1 hash of the host ID and the timestamp at the connection creation time (at a microsecond granularity). The host ID is the SHA-1 hash of the host device ID. The host device ID can be any string that uniquely identifies a host during the connection such as International Mobile Equipment Identity (IMEI) of a cell phone or the MAC address of the first interface of a laptop or a PC. The flow ID and the host ID are sent to the remote host at connection initiation, and in the rare case that the same flow ID exists at the remote host for a different connection, the remote host rejects the connection requiring the sender to retry with a new flow ID until there is no conflict.

The keep-alive duration is sent to the remote host as another op-

tion field. It is an estimated connection lifetime in seconds set by the application (e.g., dtp_setsockopt() with the KEEP_ALIVE option) before initiating the connection, and can be updated during the connection according to the application's needs. The value is negotiated by the server to limit the number of inactive connections with a very large keep-alive duration. When a host receives a packet with the keep-alive duration option, it either accepts the value by echoing it to the sender or suggests another value in the response packet until both parties agree on the same value. When the keep-alive duration is not used, DTP falls back to the normal TCP behavior and disconnects the connection after 15 retransmissions of the same packet.

## 3.3 DTP Communication

We describe the persistent data communication with DTP in three stages: connection establishment, data transmission, and teardown as shown in Figure 4(a).

### 3.3.1 Connection Establishment

To initiate a DTP connection, the sender sends a SYN packet with a flow ID, its host ID, and an optional keep-alive duration value. If the keep-alive duration is missing, the value is initialized to 0. Each host maintains a connection hash table that maps the flow ID to its remote host ID and the four tuples of the connection. If the same flow ID exists for a different connection, the receiving end responds with a RST packet to elicit a different flow ID from the sender. Otherwise, it sends a SYN+ACK with its own host ID

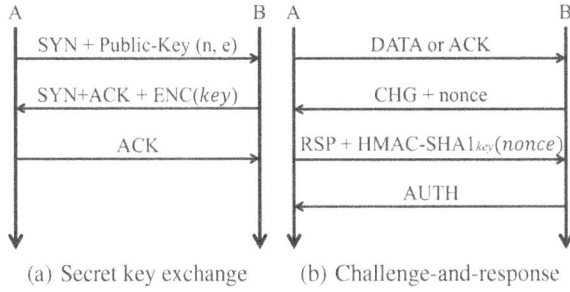

| A | B | A | B |
|---|---|---|---|
| SYN + Public-Key (n, e) → | | DATA or ACK → | |
| ← SYN+ACK + ENC(key) | | ← CHG + nonce | |
| ACK → | | RSP + HMAC-SHA1$_{key}$(nonce) → | |
| | | ← AUTH | |
| (a) Secret key exchange | | (b) Challenge-and-response | |

**Figure 5: Authentication of hosts using secret key validation**

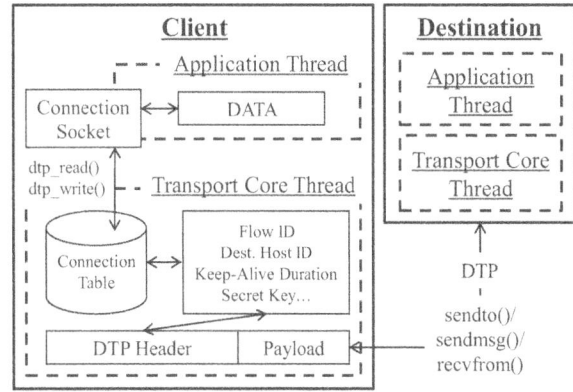

**Figure 6: Implementation of DTP**

along with the agreed flow ID and an optional keep-alive duration. To prevent connection hijacking, both parties also agree on a shared secret key, which is explained in section 3.3.4.

### 3.3.2 Data Transmission

After connection establishment, the hosts transfer the data similar to TCP. That is, slow start, sequence numbers and ACK'ing, flow and congestion control work like TCP. At network disruptions, DTP hides the failures from the application and stops data transfer until the network becomes available again. The other end host keeps sending packets until the maximum retransmission threshold is reached. It then either closes the connection or further holds the state without sending any packets, depending on the pre-defined keep-alive duration value. The network availability information can be mostly obtained from the underlying system (e.g., via the BroadcastReceiver package on Android or a netlink socket on Linux) or one can resort to periodic probing with an exponential backoff.

When the network becomes available again, DTP re-synchronizes its connection. The disrupted host sends either a normal data or an ACK packet possibly with a different IP and a port number pair, and the connection resumes after flow ID verification. When a host receives a data/ACK packet whose flow ID does not match the stored IP address and the port number, it responds with an out-of-band challenge packet with the CHG flag on. The challenge packet includes a randomly-generated nonce (8 bytes) in its payload. On receiving it, the other host replies with a response packet that has HMAC-SHA1$_{key}$(nonce) in its payload with the RSP flag on. After successful verification of the hash, the host sends an authentication packet (with the AUTH flag on), and the connection resumes (Figure 5(b)). If the verification fails, the host sends an RST packet to alert the other host to close and start a new flow.

We note that there are a few cases where the connection resumption may fail. When either host reboots for some reason (e.g., battery outage), it loses all previous connection information. If packets never arrive within the keep-alive duration, the host closes the connection on its end and notifies the application of an error (Figure 4(b)). When a host receives a packet whose flow ID does not exist in its connection table, it responds with a RST packet that has its host ID in the option field. With the host ID, the other end host checks whether the RST packet was sent by the communicating host or by another host that happens to be assigned with the same IP address of the original host after the original host left the network. If a host receives a RST with an unexpected host ID, the connection goes into the wait mode until a resumption packet from the original host arrives or the keep-alive duration expires. When both hosts change their IP addresses during a network disruption, they will end up with closing the connection after the keep-alive duration. However, this case would be uncommon in practice since most connections are between a mobile client and a server whose IP address rarely changes.

### 3.3.3 Connection Teardown

The connection can be torn down in two ways. The two hosts can explicitly close the connection by exchanging FINs and ACKs like in TCP. If the keep-alive duration expires, the host closes the connection on its end unilaterally. If the application closes the connection during the network disruption, DTP closes the connection on its end but sends a FIN to the other end when the network becomes available again within the keep-alive duration.

### 3.3.4 Shared Secret Key Exchange

To prevent connection hijacking attacks by a random host, DTP exchanges the secret key at connection setup as shown in Figure 5(a). We use the RSA algorithm here but any asymmetric key cryptographic algorithms can be used instead. The SYN packet includes the public key of the host, (n, e), in its payload, and the other end generates a secret key, encrypts it by the public key, and sends it in the payload of the SYN+ACK packet. This shared key is used to verify the ownership of the flow ID when a connection resumes after IP address change. Since the client side usually initiates the connection, the RSA decryption burden is shifted to the client side in our scheme, alleviating the load at the server. We believe that the additional work done during the secret key exchange or the challenge-and-response does not impose much overhead on the server since a modern CPU core can do more than 3,000 RSA decryptions per second [26] and heavy cryptographic operations can be easily offloaded to GPUs [27].

## 4. IMPLEMENTATION

In this section, we describe our implementation of the DTP prototype and its API library, which is designed to be compatible to that of TCP for easy migration.

### 4.1 Architecture

We implement the DTP prototype as a user-level UDP library. We choose the user-level approach for portability and ease of programming, but the more efficient kernel-level implementation would not be too hard. The DTP library spawns a "transport core" thread per application that manages the connection information and processes received packets, and the application thread provides the TCP socket-like functions to the application as shown in Figure 6. Our current prototype is compatible to TCP in terms of functionality: it implements slow start, flow and congestion control, fast re-

```
int dtp_socket(void);
int dtp_bind(int sockfd, const struct sockaddr *addr,
             socklen_t addrlen);
int dtp_connect(int sockfd, const struct sockaddr *addr,
             socklen_t addrlen);
int dtp_listen(int sockfd, int backlog);
int dtp_accept(int sockfd, struct sockaddr *addr,
             socklen_t *addrlen);
ssize_t dtp_read(int fd, void *buf, size_t count);
ssize_t dtp_write(int fd, const void *buf, size_t count);
int dtp_close(int fd);
int dtp_select(int nfds, fd_set *readfds, fd_set *writefds,
             fd_set *exceptfds, struct timeval *timeout);
int dtp_fcntl(int fd, int cmd, ... /* arg */ );
int dtp_getsockopt(int sockfd, int level, int optname,
             void *optval, socklen_t *optlen);
int dtp_setsockopt(int sockfd, int level, int optname,
             const void *optval, socklen_t optlen);
uint32_t dtp_getflowid(int sockfd);
```

**Figure 7: DTP API Functions**

transmit and recovery, timeout and retransmission, delayed ACKs, and so on.

## 4.2  DTP API Library

One of our implementation goals is to provide easy transition from TCP-based applications. Figure 7 shows the current set of DTP functions that are designed to map to a subset of TCP socket functions. dtp_socket() creates a connection context internally and returns a file descriptor to the application. Using this file descriptor, the application can connect, bind and listen on a port, accept a connection, read and write application data to the other end. A UDP socket is created internally for each connection socket (a socket through which the connection is initiated) and the server listening on a port can accept the connection by creating a UDP socket with a new port number. The mapping between UDP and DTP sockets is managed by the transport core thread. Our current implementation supports event-driven programming with dtp_select(), and we are working on implementing fork(). The number of lines of the current version is 5,283 lines in C.

Our experience with porting existing TCP servers and clients shows that it is straightforward to use the DTP library while it takes small effort to port them. We had one undergraduate student port *wget* [28] to use DTP instead of TCP socket functions. By *grep*ping socket functions and replacing them with the DTP counterparts, he could successfully port it to a DTP version in a couple of hours. It required only 19 lines of code change out of 43,372 lines of the original code. We also port a simple web server to a DTP version with the similar effort. We are currently working on porting the Apache Web server to use DTP, but it requires several function and flag options to be implemented, which is not related to actual network communication.

## 5.  EVALUATION

In this section, we compare the performance of DTP with various protocols in terms of throughput and battery consumption. In our test settings, we use a laptop with a Intel Core i7-2620M processor and 4 GB of physical memory (on linux 2.6.40) and a Nexus-S phone (on Android 2.6.35.7) as clients, and a desktop machine with a Intel Core i7-2600 CPU with 8 GB RAM (on linux 2.6.38-12) as a server.

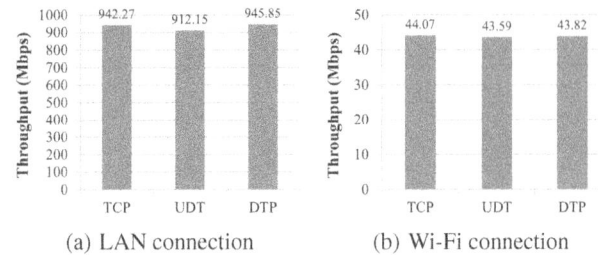

(a) LAN connection      (b) Wi-Fi connection

**Figure 8: Large-file Download Tests**

## 5.1  Microbenchmark

We first measure the base transfer throughputs between a laptop and a desktop server on a 1 Gbps LAN and on a WLAN with an 802.11n Wi-Fi AP. We compare DTP against TCP and UDT, a high-speed reliable data transport protocol based on UDP [29]. In this test, we have the laptop upload a 1 GB file to the server to saturate the LAN connection and a 100MB file for the Wi-Fi connection. As shown in Figure 8, DTP shows a comparable performance to that of TCP both on wired (945.9 Mbps vs. 942.8 Mbps) and wireless LANs (43.82 Mbps vs. 44.07 Mbps). The performance of UDT is also similar to that of TCP and DTP.

## 5.2  Performance at Network Disruptions

We now compare the DTP performance with that of TCP and the Bundle Protocol (BP), one of the representative DTN protocols [18] while the client changes its IP address when it moves to another Wi-Fi network. The clients are connected to a Wi-Fi AP whose bandwidth we limit to 3 Mbps to simulate our measured results. We use the median connection/disruption time values obtained in Section 2.1 (6.6 minutes and 7.8 minutes). If the content downloading does not complete before the disruption, we increase the next connection time by its median value. We base the file sizes by calculating the average size of the YouTube's top-viewed HD videos for the past one year (each downloaded more than 20 million times) [30], and analyze the impact of network disruptions for the file sizes of 154 MB (average file size), 77 MB (half the average), and 308 MB(double the average).

Figure 9(a) shows the throughputs between the laptop and the server. To allow fast probing of the network availability, we set the maximum backoff time of the probing packet to one second for both DTP and BP. No protocols experience a disruption for downloading the 77 MB file, and the performance is similar among all three protocols. But DTP shows 47.9% and 128.9% better performance than TCP in 154 MB and 308 MB files each. This is because TCP needs to retransmit the entire file after each network disruption, while DTP finishes downloading the files at most in one network disruption. DTP shows 3.3% to 5.2% better performance than BP because BP has extra header overhead of each bundle. While DTP shows only small performance improvement from BP, its resource consumption is much smaller than that of BP since the reference BP implementation holds the entire data in memory to prepare a bundle and creates a new primary block whenever there is a network disruption. Figure 9(b) shows the performance between the phone and the server. Similar to the previous test, DTP shows 46.9% to 122.6% better performance compared with TCP. We do not measure the BP's performance here because the reference implementation does not run on Android due to a lack of support for libraries such as oasys [31]. Another BP implementation, ByteWalla [32] runs on Android, but we find that it does not implement the "Fragment" option (Bundles are split up into multiple constituent bundles

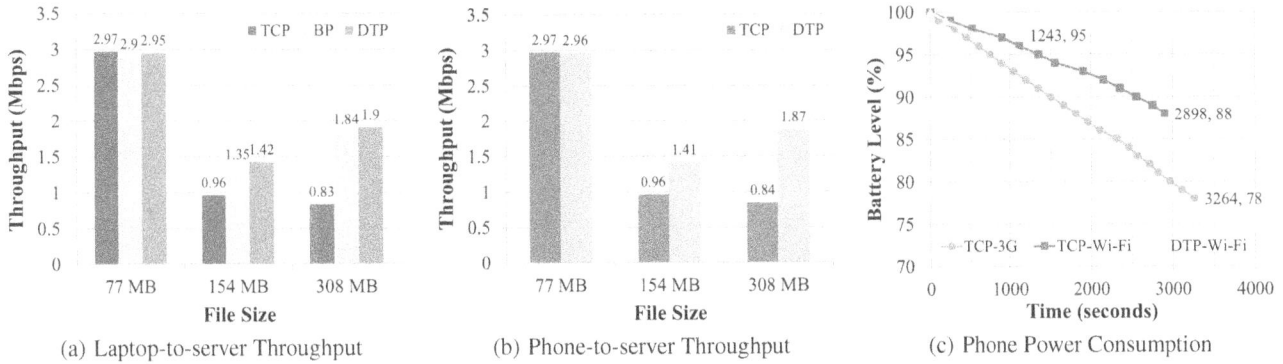

Figure 9: Throughputs and Power Consumption during Network Disruptions

at network disruptions.) [33], and cannot resume data transfer after network disruptions.

## 5.3  Power Consumption

Figure 9(c) shows the power consumption of Nexus S as we run the tests with the 308 MB file from the previous section. We also measure the power consumption of transferring the same file via TCP using the 3G network *without* disruptions (TCP-3G). TCP-3G shows a rapid decrease in battery power since it consumes more energy during data transfer than Wi-Fi as in [34]. Also, its throughput is the worst (0.78 Mbps) and finishes the last even without network disruptions. We find that DTP-Wi-Fi consumes 58.3% and 77.3% less power compared with that of TCP-Wi-Fi and TCP-3G while it finishes downloading 1,655 and 2,021 seconds earlier.

## 6.  RELATED WORKS

There have been several approaches to support mobility for IP networks. Mobile IP [35] exports a fixed home address through which external hosts communicate regardless of the actual network attachment of the mobile host. When the mobile host leaves its home network, the home agent relays the IP packets arriving at the home address to the *care-of-address* (e.g., real address) of the mobile host. In contrast, DTP does not require a home agent, nor it needs to relay packets, which would produce better packet routing between the two ends.

DTP is similar to Migrate TCP option [36] in that both enable connection reuse for IP address change. But DTP is more friendly to 3G/Wi-Fi offloading environments since it allows the applications to set the disruption delay much larger than the maximum segment lifetime (MSL) DTP bears the similarity with i3 [37] and HIP [38] in that they support mobility by separating the host identity and its network location. Unlike DTP, both require additional infrastructure support such as a DHT network and the DNS.

DTP supports the concept of delay-tolerant networking [9] into Wi-Fi data offloading. Previous DTN protocols such as Bundle Protocol (BP) [18] and Licklider Transmission Protocol (LTP) [15–17] assume more challenged networks with high delays and packet losses whereas DTP is geared towards mostly stable networks but with frequent disruptions. Exploiting the fact, DTP supports reliable transfer on an IP packet level without an additional layer that wraps the content into bundle blocks as in BP. Besides, the current BP works only for the content whose size is already known prior to transmission, whereas DTP allows users to watch a streaming video without interruption even when she moves from a Wi-Fi network to 3G or vice versa. LTP is designed to reliably transfer the data mostly in dedicated networks with very high RTTs

(e.g., deep space) and does not consider typical TCP issues such as flow and congestion control in shared networks. For this reason, the current LTP implementation uses pre-defined parameters (e.g., window size) before the communication initiates [39]. In contrast, DTP strives to conform to TCP to be fair to other competing flows, allowing the co-existence of heterogeneous networking devices.

## 7.  DISCUSSION

In this section, we discuss some of the issues that were not addressed in this paper and consider an extended offloading framework for our future work.

**State Explosion Attacks:**  In malicious environments, an attacker can instruct zombie hosts to create many DTP connections with a long keep-alive duration on a target server. While we design DTP to have a very small memory footprint per connection and allow the server to limit the keep-alive duration value to specifically guard against this attack, the application sometimes has to maintain a large buffer per request. One such scenario is that the client sends a large-file request and goes offline immediately afterwards. However, we note that this sort of attack is not unique to DTP but can be launched on any TCP-based servers. One defense approach is for the server to detect suspicious requests by careful resource accounting [40] and dynamically reset the keep-alive durations when it is suspected to be under attack. We plan to explore this issue further in the future.

**ISP-driven Offloading Servers:**  Using DTP, mobile ISPs may further exploit Wi-Fi offloading for efficient network resource utilization. One example is that mobile ISPs provide a DTP cloud storage server that runs an application protocol multiplexer. In this scenario, the multiplexer translates DTP connections from mobile hosts to TCP connections to the target server and vice versa. For instance, a client can send emails to the cloud server using DTP-based SMTP, and the cloud server relays them to the destination using TCP. Or the cloud server can receive and store a podcast video from a TCP-based postcast server and pushes it onto the mobile phone using DTP. This would not only provide an incremental deployment path of DTP, but also allow the mobile ISPs to spread bandwidth consumption across the time axis similar to SmartGrid [41]. We are currently working on a cloud storage server for mobile ISPs.

## 8.  CONCLUSION

While many works have shown the effectiveness of Wi-Fi mobile data offloading, there has not been a practical data delivery mechanism to support it. We propose DTP, a disruption-tolerant reliable transport layer protocol, which allows seamless switching between

3G and Wi-Fi networks on the same connection for mobile applications. We design it to easily migrate existing applications to transparently recover from network disruptions, with little performance degradation from that of TCP. Our evaluation shows that DTP is promising with the great potential to reduce 3G network usage as well as the battery consumption.

# 9. ACKNOWLEDGEMENTS

We thank anonymous reviewers for their insightful comments. We also thank Yung Yi and Song Chong for lively discussion about the role of the disruption-tolerant transport layer. This work was supported by the KCC (Korea Communications Commission), South Korea, under the R&D program supervised by the KCA (Korea Communications Agency), KCA-2011-11913-05004.

# 10. REFERENCES

[1] SEO Updates. Mobile vs Desktop Internet Usage Stats 2011, 2011. http://www.seodailyupdates.com/2011/06/mobile-vs-desktop-internet-usage-stats.html.

[2] CISCO. Cisco Visual Networking Index: Global Mobile Data Traffic Forecast Update, 2010-2015. Technical report, 2011.

[3] J. M. Chapin and W. H. Lehr. Mobile Broadband Growth, Spectrum Scarcity, and Sustainable Competition. In *Proceedings of The 39th Research Conference on Communication, Information and Internet Policy*, 2011.

[4] J. Mitola III and G. Q. Maguire Jr. Cognitive Radio: Making Software Radios More Personal. *IEEE Personal Communications*, 6(4):13–18, 1999.

[5] A. Balasubramanian, R. Mahajan, and A. Venkataramani. Augmenting Mobile 3G Using WiFi. In *Proceedings of ACM MobiSys*, 2010.

[6] K. Lee, I. Rhee, J. Lee, S. Chong, and Y. Yi. Mobile Data Offloading: How Much Can WiFi Deliver? In *Proceedings of ACM CoNEXT*, 2010.

[7] J. Scott, P. Hui, J. Crowcroft, and C. Diot. Haggle: A Networking Architecture Designed Around Mobile Users. In *Proceedings of IFIP WONS*, 2006.

[8] A. Chaintreau, P. Hui, J. Crowcroft, C. Diot, R. Gass, and J. Scott. Impact of Human Mobility on the Design of Opportunistic Forwarding Algorithms. In *Proceedings of IEEE INFOCOM*, 2006.

[9] K. Fall. A Delay-Tolerant Network Architecture for Challenged Internets. In *Proceedings of ACM SIGCOMM*, 2003.

[10] Podcast. http://www.apple.com/itunes/podcasts/.

[11] DoggCatcher. http://www.doggcatcher.com/.

[12] TubeMate. http://tubemate.tistory.com/.

[13] Dropbox. https://www.dropbox.com/.

[14] Google/Ipsos OTC MediaCT. The Mobile Movement Study, 2011.

[15] S. Burleigh, M. Ramadas, and S. Farrell. Licklider Transmission Protocol - Motivation. RFC 5325, IETF, 2008.

[16] M. Ramadas, S. Burleigh, and S. Farrell. Licklider Transmission Protocol - Specification. RFC 5326, IETF, 2008.

[17] S. Farrell, M. Ramadas, and S. Burleigh. Licklider Transmission Protocol - Security Extensions. RFC 5327, IETF, 2008.

[18] K. Scott and S. Burleigh. Bundle Protocol Specification. RFC 5050, IETF, 2007.

[19] KT. olleh WiFi zone Finder. http://zone.wifi.olleh.com/en/index.action.

[20] Android Market. https://market.android.com/.

[21] MapDroyd. http://www.mapdroyd.com/.

[22] Beyondpod. http://www.beyondpod.mobi/android/index.htm.

[23] Google Listen. https://market.android.com/details?id=com.google.android.apps.listen/.

[24] Winamp. http://www.winamp.com/android/.

[25] R. Braden. Requirements for Internet Hosts - Communication Layers. RFC 1122, IETF, 1989.

[26] Michael E. Kounavis, Xiaozhu Kang, Ken Grewal, Mathew Eszenyi, Shay Gueron, and David Durham. Encrypting the internet. In *Proceedings of ACM SIGCOMM*, 2010.

[27] K. Jang, S. Han, S. Han, S. Moon, and K. Park. SSLShader: Cheap SSL Acceleration with Commodity Processors. In *Proceedings of USENIX Symposium on Networked Systems Design and Implementation (NSDI)*, 2011.

[28] GNU wget. http://www.gnu.org/s/wget/.

[29] Y. Gu and R. L. Grossman. UDT: UDP-based Data Transfer for High-Speed Wide Area Networks. *Computer Networks (Elsevier)*, 51(7), 2007.

[30] YouTube. http://www.youtube.com.

[31] Oasys Documentation Library. http://www.omgeo.com/page/productdocumentation/?var1=oasys/.

[32] R. Yanggratoke, A. Azfar, M. J. P. Marval, and S. Ahmed. Delay Tolerant Network on Android Phones: Implementation Issues and Performance Measurements. *Journal of Communications*, 6, 2011.

[33] V. Cerf, S. Burleigh, L. Torgerson, R.Durst, K. Scott, K. Fall, and H. Weiss. Delay-Tolerant Networking Architecture. RFC 4838, IETF, 2007.

[34] N. Balasubramanian, A. Balasubramanian, and A. Venkataramani. Energy Consumption in Mobile Phones: A Measurement Study and Implications for Network Applications. In *Proceedings of ACM Internet Measurement Conference (IMC)*, 2009.

[35] A. Myles, D. Johnson, and C. Perkins. A mobile host protocol supporting route optimization and authentication. *IEEE Journal on Selected Areas in Communications*, 13(5), 1995.

[36] A. C. Snoeren and H. Balakrishnan. An end-to-end approach to host mobility. In *Proceedings of ACM MOBICOM*, pages 155–166, 2000.

[37] I. Stoica, D. Adkins, S. Zhuang, S. Shenker, and S. Surana. Internet Indirection Infrastructure. 2002.

[38] R. Moskowitz and P. Nikander. Host Identity Protocol Architecture. RFC 4423, IETF, 2006.

[39] S. Farrell, V. Cahill, D. Geraghty, I. Humphreys, and P. McDonald. When TCP Breaks: Delay- and Disruption-Tolerant Networking. *IEEE Internet Computing*, 10(4), 2006.

[40] Xiaohu Qie, Ruoming Pang, and Larry Peterson. Defensive Programming: Using an Annotation Toolkit to Build DOS-Resistant Software. In *Proceedings of USENIX Symposium on Operating Systems Design and Implementation (OSDI)*, 2002.

[41] R. E. Brown. Impact of Smart Grid on distribution system design. IEEE, Power and Energy Society general Meeting, 2008.

# QuickSilver: Application-driven Inter- and Intra-cluster Communication in VANETs

Riccardo Crepaldi
University of Illinois
at Urbana Champaign
rcrepal2@illinois.edu

Mehedi Bakht
University of Illinois
at Urbana Champaign
mbakht2@illinois.edu

Robin Kravets
University of Illinois
at Urbana Champaign
rhk@illinois.edu

## ABSTRACT

Support for efficient vehicle-to-vehicle communication is increasingly more important with the emergence of newer vehicles equipped with one or more wireless interfaces. While the applications aimed at such networks range from car-to-car chats to sharing dynamic map data, current approaches to inter-vehicular networking have been designed based on either node-centric or content-centric communication, but not both. The challenge for a comprehensive solution arises from the high mobility of nodes as well as the heterogeneity of contact patterns. In this paper, we propose QuickSilver, a system architecture that meets this challenge by leveraging an intrinsic characteristic of vehicular networks - clustering. By being aware of such clustering, Quicksilver enables a seamless integration of two networking paradigms, one for node-centric communication between members of the same cluster, the other for effective content dissemination and exchange during short cluster-to-cluster contacts. To achieve this goal efficiently, Quicksilver employs a novel combination of channel management and light-weight clustering that enables detection and management of headless, multi-hop clusters at very low cost. Evaluation results show that Quicksilver enables close-to-optimal performance when nodes across different clusters communicate.

## Categories and Subject Descriptors

C.2.1 [**Computer-Communication Networks**]: Network Architecture and Design—*Distributed networks,Store and forward networks*; C.2.2 [**Computer-Communication Networks**]: Network Protocols

## Keywords

VANETs, Clustering, Vehicular Distributed System

## 1. INTRODUCTION

Vehicular networks (VANETs) turn the tables on mobile social networks, where the devices are carrying the people instead of the people carrying the devices. The promise of vehicular networks has led many researchers to design systems and protocols for enabling wireless communication between vehicles. Since it is still under discussion what type of applications need to be supported in vehicular networks, many general solutions have been proposed based on ad hoc [1, 2] or DTN-based [3, 4] routing. However, these solutions focus on point-to-point communication between vehicles, which may not always be the best communication paradigm for vehicular networks. Additionally, they overlook the dynamic characteristics of the networks, missing out on important solutions and optimizations. On the other hand, many solutions have been proposed for more specific scenarios that focus on a vehicle's ability to find an open or accessible access point [5, 6]. These solutions only support applications that need access to the Internet and also ignore the available networking and data resources found locally around a given vehicle.

In reality, vehicular networks are a special form of an opportunistic social network with highly constrained mobility rules that typically result in heavy clustering in the network [7]. The concept of clustering is in no way new and it has been used as a means for providing optimized communication in both DTN protocols to improve forwarding decisions [8] and in last-hop style vehicular networks to support extended connectivity to two or more hops around an access point. However, if one looks at clustering not as an optimization possibility, but as an intrinsic characteristic of vehicular networks, clustering not only shapes the network topology, but also results in a set of applications with requirements and communication patterns that are so different that clearly they cannot be supported efficiently by a single communication paradigm.

The distinct mobility patterns and sheer scale of VANETs result in geographically wide-spread networks with unique contact opportunities between the diverse nodes. Essentially, related and non-related clusters of vehicles may travel together for long periods of time, briefly encountering other random nodes and clusters that are traveling in different directions. This clustered mobility pattern results in interesting data exchange patterns in VANETs. Although the people in the vehicles inside a cluster may not actually know each other, they may still want to share music, videos, or even just chat with each other because they are going in the same direction like old CB users. Since clusters are somewhat stable, *intra-cluster* communication should support these *node-centric* applications with relatively stable, good quality links to specific vehicles. However, such support is no longer necessary, nor even possible, beyond the border of a cluster. Therefore, the focus of *inter-cluster* communication should not be on the endpoints of the communication but instead on supporting *content-centric* applications that aim to distribute shared data, such as dynamic maps and participatory sensor data, with many unknown nodes across the network.

To cover the whole scope of VANETs, many researchers have classified VANETs as Delay Tolerant Networks (DTNs) with dynamic mobility models, network partitioning and the natural formation of clusters. Since energy is not a severe constraint in moving vehicles, DTN-based VANET routing solutions have focused on supporting node-to-node communication that is robust to network partitioning and intermittent connectivity. On a large scale, it is not clear that such node-centric support is needed when nodes are far apart. Instead, support for geographically distributed nodes may be better off relying on 3G- or hot-spot-based solutions. However, content-centric applications, which rely on probabilistic dissemination of data between clusters, match well with existing DTN-based communication. While probabilistic replication mechanism can be used, the main challenge comes from establishing efficient inter-cluster communication due to the short connection times and contention for the channel caused by the clustering of vehicles. On the local scale, clustered nodes within a smaller geographic area may want to communicate with each other. However, in this case, simpler ad hoc-based routing solutions may be sufficient. While some existing solutions have leveraged the natural clustering in VANETs to optimize message delivery, they have not considered the important impact of such clustering on the application space.

While intra- and inter-cluster communication have very diverse requirements, a complete VANET solution must support both. To this end, we propose QuickSilver, a system architecture built around a novel, passive cluster management mechanism that does not require the use of cluster ids or other explicit clustering mechanisms. This clustering mechanism enables both intra-cluster networking support for node-centric applications, as well as inter-cluster networking support for content-based applications. For the latter, nodes within a cluster implicitly cooperate to establish as many contention-free inter-cluster communication channels, without requiring explicit channel allocation mechanisms. The goal of QuickSilver is the efficient the use of the available resources to guarantee that no harmful competition over the limited channel bandwidth is engaged. Our evaluation shows that QuickSilver's clustering mechanism can maintain good consistency for cluster memberships, despite high mobility and its distributed nature. Additionally, we show that QuickSilver is able to establish multi-link inter-cluster connections that perform significantly better than the existing hierarchical approaches, and in some scenarios reach 95% of the performance of an optimal, centralized solution, without incurring any control overhead. In the rest of this paper, we discuss the challenges of such a comprehensive solution and present the design and evaluation of QuickSilver.

# 2. VANET ECO-SYSTEMS

The formation of clusters in vehicular networks has been observed in different scenarios. For example, the analysis of mobility traces from 600 taxis over an 8 hour period on a working day in San Francisco [7], showed that while 40% of the cabs were sparsely connected, the remaining 60% were part of clusters of more than 20 vehicles, whose routes periodically intersect generating intercluster contacts. Using short to mid-range transceivers, vehicles can communicate with each other to distances up to 200m [9]. This distance, which is certainly larger than the average distance between vehicles in a cluster, is also likely to be shorter than the length of the cluster itself. This implies that from a networking point of view, clusters can be seen as partitions of the network with a number of stable, multi-path links among their nodes.The duration of a vehicle's membership to a specific cluster is influenced by the speed at which vehicles move, and the number of different routes available. In urban environments, low speed ensures a certain duration of the average membership duration, while on high-

ways the same property is a consequence of the distance between exits.

## 2.1 Cluster-Aware Communication

Given this clustering and the diversity of applications in VANETs, many different kinds of applications may coexist, each characterized by different requirements and goals. However, these applications can be divided into two c'ategories: node-centric and content centric. Node-centric applications focus on providing a distinct communication channel between two or more end hosts. For example, nodes A and B in Figure 1(a) are participating in a voice conversation, while nodes X and Y are using a real-time instant messaging application. each node must be able to learn about about each other's existence to be able to establish a good quality and long lived link for bounded-delay communication, both of which limit node-centric communication to within a cluster. Essentially, except for scenarios where nodes follow predetermined routes (i.e., DieselNet [10]), it is extremely unlikely to meet a specific user during an inter-cluster contact, and even when this happens, the contacts are too rare and short-lived to support node-centric networking.

On the other hand, content-centric applications focus on data that lives in the network [11, 12, 13], where nodes are the relays that spread the data through the network. This type of data is typically spread through the network based on nodes' interests in a particular type or class of data. For example, two nodes in one cluster in Figure 2(b)) are expressing interest in some type of music, while one node in the other cluster is sharing songs. If a path exists between the content provider and the consumers, the data can be shared among the nodes. In this case, the identity of the nodeqs is completely irrelevant, only the interests they express matter. This kind of communication can cross the boundaries of a cluster. Even with short-lived inter-cluster connections, applications can benefit from receiving as much data as possible from the other cluster.

Content-centric applications can also be used to collect and disseminate information about the environment [11]. The vehicles in Figure 1(c) are equipped with a large number of sensors to collect environmental information. This information can be used to augment a map with live data, useful for example for route planning. In this type of application, the data is shared among vehicles that have been traveling together, and each of them is part of a distributed storage. When a cluster traveling in the opposite direction is encountered, it is irrelevant which nodes share the information, as long as it is transferred from one cluster to the other, where it can be distributed to all nodes even when the cluster contact has expired.

## 2.2 Cluster Contacts

Contact opportunities between nodes that belong to different clusters are generally short. Given two vehicles traveling on an highway in opposite directions and approaching each other at speeds $v_a$ and $v_b$, contact is possible only while their relative distance is smaller than the wireless coverage range $R$. The contact duration in this case is $T_{cont} = 2R/(v_a + v_b)$. Assuming a coverage range of $200m$ and a speed of $15m/s$ for both cars, a contact between two vehicles can last approximately $15s$, a very short time for any significant data exchange. However, cluster contact opportunities last from when the first contact between two vehicles is possible, until the distance between the last pair of cars becomes larger than $R$. For two clusters traveling in opposite directions, with velocities $v_a$ and $v_b$, and cluster lengths $L_a$ and $L_b$, both larger than $R$ (i.e., multi-hop clusters), the overall duration of the cluster contact is $T_{c\_cont} = \lceil L_a/2R \rceil \cdot T_{cont}$, which could allow a significant

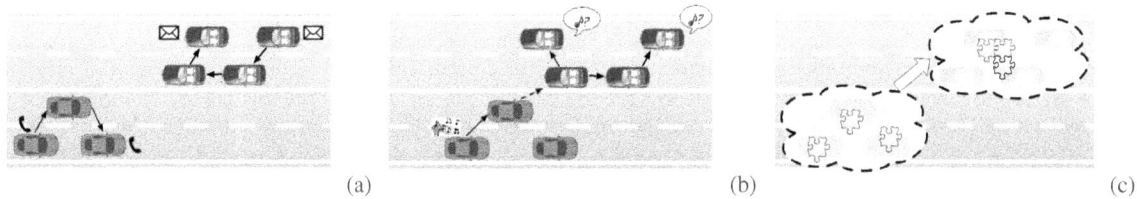

Figure 1: The types of communication that characterize a VANET: *intra-cluster* (a), *inter-cluster* (b), *cluster-to-cluster* (c)

amount of data exchange among the two clusters if the communication channels are managed wisely.

## 2.3 Cluster Management

Understanding the network topology and its organization in clusters is necessary to define the boundaries of intra-cluster connectivity and determine which links belong to an inter-cluster contact. Traditional cluster management approaches are the basis of communications systems such as Bluetooth or ZigBee, where clustering enables coordinated, hierarchical communication to save energy and increase security. A similar hierarchical structure, based on the election of a cluster head (CH), has been considered in previous research work for VANETs [14, 15, 16]. For example, clustering can be used to coordinate medium access to avoid collisions [17] by allowing the CH to acquire information about the cluster members and then coordinates their accesses to the share medium using a TDMA approach. All of these types of approaches view of clusters in VANETs as single-hop subnetworks concentrated around the cluster head. Since membership typically changes rapidly due to the nodes' independent mobility patterns in mobile networks like VANETs, multi-hop clusters are typically dismissed due to the amount of overhead required for electing a cluster head and maintaining cluster membership information. However, the use of multi-hop links has been proposed in VANETs without explicit clustering. For example, CASCADE [6] is an access point-based approach where vehicles traveling in the same direction forward data through each other to an access point. However, approaches like CASCADE are not able to leverage inter-cluster contacts since they do not have any concept of a cluster.

## 2.4 Inter- and Intra-Cluster Routing

Given the diversity of application requirements and the dynamic mobility in the network, routing solutions for VANETs must be able to integrate both intra-cluster, node-based routing with bounded delays and inter-cluster content-based routing with quick and efficient data dissemination. Traditional ad hoc routing protocols (i.e., DSR [18], AODV[19]) can only need slight modifications to provide the bounded delays for intra-cluster communication, but fail dramatically to to support the opportunistic inter-cluster networking. On the other hand, store-carry-and-forward DTN protocols (i.e., Prophet [20], Spray and Focus [4]) as well as protocols specifically designed for data-centric VANET applications (i.e., Locus [11]) are resilient to partitioning, but cannot effectively support node-centric applications. Essentially, a one-size-fit-all protocol cannot satisfy the requirements of both intra- and inter-cluster networking. Although protocols have been designed with one or the other pattern in mind, it is unclear how they would or should interact in the same system. If the two protocols compete in an uncoordinated manner to access the shared resources (i.e., available channels and bandwidth) on a node, their combined performance might be severely degraded. Only with a careful system design that is aware of the network topology and cluster structure and coor-

dinates the use of resources can the full potential of VANETs be achieved.

## 3. QUICKSILVER

While node-centric and content-centric communication in VANETs have very different network requirements, the key feature of clusters enables a symbiotic solution. In this section, we present Quick-Silver, a system architecture that leverages the clustered topology for supporting the different types of communications in VANETs. Cluster detection is pivotal to tracking the boundaries inside which node-centric communications can happen. When nodes know what other nodes are in the same cluster, they can cooperate to route intra-cluster traffic, and at the same time concentrate their effort on detecting inter-cluster contacts and maximize the amount of data that is exchanged between two clusters.

To enable effective and efficient communication in VANETs, the QuickSilver design is inspired by the following principles. First, QuickSilver is *completely distributed*. No infrastructure is needed for communication. Second, QuickSilver employs *lightweight clustering*, where clusters form and behave in an uncoordinated manner without requiring a cluster ID. There are no cluster coordinators in QuickSilver. Therefore, each node builds its own view of the cluster, which might be partial and differ from node to node. Finally, QuickSilver supports *cluster-aware communication*, where intra-cluster networking leverages the fast delivery guaranteed by stable paths inside a cluster and inter-cluster networking manges the cluster-to-cluster channels to opportunistically support content-centric communication.

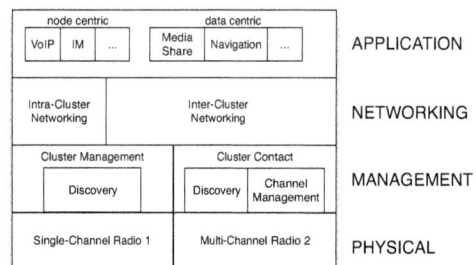

Figure 2: QuickSilver system architecture

## 3.1 System Architecture

The general architecture of QuickSilver is shown in Figure 2. Similar to [3], QuickSilver utilizes two radio interfaces to support concurrent communication using orthogonal channels. QuickSilver is not tied to a specific wireless standard. It can use generic 802.11a/g/n, or the 802.11p standard designed for Digital Short Range Communications (DSRC). This dual radio setup allows the vehicles to maintain their intra-cluster connectivity, and at the same time look for inter-cluster contact opportunities. However, the resource allocation policy in QuickSilver is different from that of other clustering based approaches that prioritize intra-cluster communications by assigning more channels for it and reserving only one channel for data exchange between clusters. Since intra-cluster communication is typically for longer-lived connections, QuickSilver uses one radio and a fixed channel for all intra-cluster communication. Therefore intra-cluster communication is easier to manage and requires no coordination, enabling better support for node-centric communication. Additionally, it leaves more channels available for the short-lived inter-cluster communication, increasing the bandwidth achievable during an inter-cluster contact.

The *management* layer, which includes *Cluster Management* and *Cluster Contact* components, is the heart of QuickSilver's clustering. The goal of the cluster management component of QuickSilver is to discover neighbors with stable links, and build Cluster Membership Lists (CML) in each node. Cluster formation is based on the rationale that nodes traveling together in the same direction and at similar speeds experience links that are long-lived enough to support multi-hop path-based intra-cluster routing. Therefore, only nodes connected through such *stable links* are considered members of the same cluster.

Cluster membership is very dynamic, with continuous joins and leaves. Keeping a consistent view of the cluster requires a mechanism to maintain an up-to-date CML in each node, and of course this property can be satisfied only with a certain amount of overhead. To minimize cluster-wide negotiation and overhead, QuickSilver does not try to enforce a globally consistent view on the cluster. Rather, each node forms its own view of its cluster and maintains a CML of those it believes to be members of its cluster.

Using the inter-cluster radio interface, the *cluster contact* looks for the short-lived contacts with other clusters traveling in different directions. The ultimate goal of QuickSilver in this respect is to build links across the two clusters that support to the formation of a *fat pipe* between the two clusters. This pipe is opportunistically established to maximize the data exchanged between the two clusters. Independently from which nodes are transmitting, the goal is to establish the largest pipe between the clusters and maximize its utilization, using non-interfering ad hoc links between pairs of nodes belonging to the two clusters.

Finally, the *networking* layer implements the multi-hop delivery for both intra-cluster and the inter-cluster communication. The former only uses links established through the single-channel interface, while the latter uses both interfaces in the data path between content providers and consumers.

## 3.2 Cluster Management

The construction and maintenance of CMLs in QuickSilver is a two-step procedure: first, a node identifies its stable neighbors, then this local topology is distributed so that each node can update its CML with cluster members that are more than one hop apart. A downside of this approach is that nodes in the same cluster might temporarily have different views of the cluster memberships (i.e., incomplete lists or stale information, caused by recent changes in the cluster topology), and the system must be robust to such inconsistencies. However, given the opportunistic nature of communication in VANETs, such inconsistencies should have little or no effect.

### Identifying stable neighbors

Neighbor discovery in opportunistic networks is generally supported by beacons or simple keep-alive message. QuickSilver takes advantage of these existing neighbor discovery mechanisms for evaluating the stability of a connection to a given neighbor. Each node broadcasts hello messages with a fixed period $T_h$. These mes-

sages contain the source node ID, and a sequence number, incremented for each new message. If a node receives $th_{join}$ sequential hello messages from a neighbor, it marks that neighbor as stable. If $th_{leave}$ periods expire without receiving a hello message from a neighbor, the neighbor is removed from the neighborhood table. The identification of these one-hop stable links is central to the notion of clusters in QuickSilver.

### Building the Cluster Membership List (CML)

Any node that becomes a stable neighbor of a node is immediately inserted into the node's CML. To identify cluster members that are multiple hops away, each node periodically broadcasts a c_hello message to all of its stable neighbors. The period $T_c$ is larger than $T_h$. Relaying nodes attach their IDs to the message and forward it over all stable links. Any node receiving a c_hello message immediately puts the original sender of the message and all the nodes that relayed the message in their CML.

Since intra-cluster routing requires the use of broadcast messages, which will be describe in Section 3.4, QuickSilver leverages these broadcast messages to enable better and more up-to-date cluster management. When a node initiates one of these broadcasts, the message is also used as a c_hello, and the relative timer is reset. The relaying nodes piggyback their IDs onto the control message but their timers are unaffected. It is important to note that the use of adding node IDs to every broadcast message can be expensive in large networks. However, these techniques are only used within clusters, which are expected to be relatively small in size, and so will not incur as much overhead as if used network-wide.

While stable neighbors are always part of the CML, destabilization of a stable neighbor does not immediately lead to its removal from the list. The reason is that the node does not know if the single-hop link, which has now become unstable, was the only stable path to that node. A node is removed from the CML only if more than $th_{c\_leave}$ cluster advertisement periods $T_c$ pass after its latest insertion in the set (using either c_hello or any cluster-wide broadcast message).

A simple analysis gives us an upper bound for the time it takes nodes in a cluster to reach a consistent view on cluster membership (i.e., the CML on each node is identical). When a node joins a cluster, that node must have at least one stable link with a neighbor in the cluster, resulting in the insertion of the new node in the CML of its one-hop neighbors. When the proper timer on the node expires, a c_hello message containing the new node ID is propagated throughout the cluster and each cluster member updates their CML accordingly. Assuming that MAC retransmission policies and the redundancy introduced by multiple routes cancels the impact of packet loss and that the time lapse from when a node creates a message and one of its neighbors receives it is $T_{tx}$, the largest time required for a new node to be present in all cluster members' CMLs is: $T_{cons\_join}^{max} = (th_{join} \cdot T_h) + T_c + T_{tx} \cdot \max(n_{hops})$, where the first term captures the time necessary to stabilize the local link, the second is the maximum time necessary for the c_hello timer to expire after the link is stabilized (in the worst case, the times could have fired right before the new link is stabilized), and the third one is the propagation time necessary for the c_hello message to travel to the farthest node in the cluster. Similarly, the maximum time necessary to reach a consistent view after a node leaves the cluster is simply the timeout of a CML entry, which is: $T_{cons\_leave}^{max} = T_c \cdot th_{leave}$.

Given these limits, to achieve a consistent view of the cluster, $T_c$, $T_h$, $th_{join}$ and $th_{leave}$ must be chosen accordingly to the cluster dynamics. The right tradeoff between system responsiveness and overhead must be found. Additionally, it should be noted that a too short $T_h$ or a too low $th_{join}$ would cause short-lived links to be erroneously considered stable. This is a worst case analysis, and in most cases a consistent view might be reached with less strict requirements. However, although QuickSilver's performance would benefit from a consistent view, the system design is robust to inconsistencies, and the timer values and thresholds should be chosen trying to limit the control overhead they generate.

## 3.3 Cluster contacts

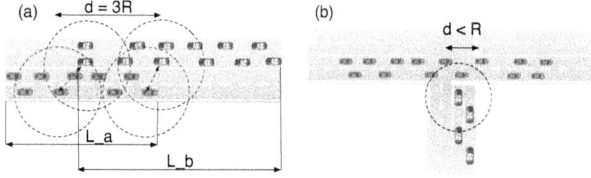

**Figure 3: Single-channel connection for inter-cluster contacts.**

Although there are mechanisms that can be used to establish contention free communication over shared channels, such as TDMA, they incur significant overhead due to synchronization and coordination. The mobility of clusters poses an additional obstacle to centralized, coordinated solutions: the connection between clusters can be maintained for the whole $T_{c\_cont}$ period, but the single links that keep it alive change constantly as nodes move and change neighbors. For this reason, QuickSilver adopts a technique that is completely distributed and opportunistic, an alternative that does not require control overhead.

The fat pipe between two clusters is built on $N_c$ links, where $N_c$ is the number of available orthogonal channels, excluding the one used for intra-cluster communication. Each link is locally used by a single pair of nodes, one per contention area in a cluster, so that no contention is necessary and the channel utilization is maximized, as long as the nodes' queues are not empty. These nodes are *gateways* through which the traffic from their cluster members can be forwarded to the other cluster. Once they reserve the channel they are the only nodes authorized to transmit on that channel in their coverage range. As shown in Figure 3(a), if $N_c = 1$ (i.e., only one channel is assigned to inter-cluster communication), when two clusters are traveling in opposite directions like on an highway, if they overlap for a length $d$, a number of links $N_L^S \in [\lceil d/R \rceil, \lceil d/2R \rceil]$ links can be established. When two clusters meet at a crossing, as shown in Figure 3(b), $d < R$, thus only one contention free link can be established. This is equivalent to having single hop clusters where only the cluster head is authorized to perform inter-cluster communication [3, 15]. If $N_c > 1$, the number of possible links becomes $N_L = N_c \cdot N_L^s$ for the highway scenario and $N_L = N_c$ in the crossing scenario. Building the single links involves two challenges: first nodes must detect a contact opportunity, then they must select an available channel and become its owners.

In all mobile scenarios, where contacts can be short lived, delay in discovery reduces the contact utilization enough to degrade the performance. In general, two factors influence the delay in discovering a contact. The first one is energy savings mechanisms that force the radio to sleep for long periods of time. In this case, even if two nodes are in communication range, they must wait until both radios are on at the same time to initiate discovery. Fortunately, QuickSilver does not require an aggressive power-saving mode, given that the energy consumed by the radio is negligible compared to the rest of the system when the vehicle is running. However, QuickSilver suffers from another source of delay in contact discovery: the presence of multiple channels. For a successful

discovery, two nodes must be tuned on the same channel at the same time. One approach to meet this requirement is the agreement on a specific inter-cluster control channel as in [3], where all nodes not currently part of an active link are advertising their presence. When two nodes meet on the control channel, they exchange control packets to agree on a different channel on which they should both switch to and communicate. This approach causes the waste of one channel for control packets. Additionally, unless an omniscient cluster head has full knowledge of which nodes are using which channels, as well as their interference range, there can be no guarantee that the negotiation phase brings the two nodes to a free channel. This challenge is amplified by the multi-hop nature of clusters in QuickSilver.

QuickSilver takes a completely distributed approach that avoids wasting one channel for control traffic. The principle on which QuickSilver is based is very simple: each node channel hops following a randomized permutation of the $N_c$ channels, sending a `disc` message if the channel is free. The message contains the node's CML. If a node from a different cluster is on the same channel, it responds to the message and the two nodes form a inter-cluster link. Since consistent cluster IDs are not possible in Quick-Silver's lightweight clustering, nodes determine if they belong to different clusters by comparing their CMLs and verifying if they differ for a fraction larger than $th_{CML} \in [0; 1]$.

The randomized channel hopping sequence must be carefully chosen to guarantee that two nodes will meet on the same channel at least once within a bounded time, and that they will meet on a free channel, if any is available. QuickSilver's sequence generation algorithm is inspired by SSCH [21], where each node selects a seed $s \in [1, N_c]$ and a starting channel $x_0$. When $N_c$ is a prime, the i-th element of the hopping sequence is: $x_i = (x_{i-1} + s) \bmod N_c$. Two nodes with a different $s$ are guaranteed to meet on one of the channels in at most $N_c$ hops. However, if the channel is busy the nodes stay silent, thus they do not discover each other despite being on the same channel. QuickSilver introduces a novel sequence algorithm: $x_i = ((x_{j-1} + s) \bmod N_c) + k \bmod N_c$, where $j = i \bmod N_c$ and $k = \left\lfloor \frac{i}{N_c} \right\rfloor$. This algorithm guarantees that two nodes with a different $s$ will meet at least once on *each channel* after $N_c^2$ hops, as shown in Figure 4, so that they can establish a link unless all channels are busy. If $N_c$ is not prime, the same property can be achieved by choosing a prime number $M > N_c$ and modifying the hopping sequence using a modulus reduction: $x_i = ((x_{j-1} + s) \bmod M) \bmod N_c + k \bmod N_c$, where $j = i \bmod M$, $k = \left\lfloor \frac{i}{M} \right\rfloor$, and the number of hops required for a pair of nodes to meet on each channel becomes $M \cdot N_c$.

If two nodes have the same seed, and they start from a different channel, they will follow each other without ever meeting. To prevent this from happening, SSCH introduces parity slots, but unfortunately this requires synchronization among nodes to guarantee that the parity slots happen at the same time, which would be impractical in a large distributed vehicular network. Since the goal during an inter-cluster contact is just to build a link with any node from the other cluster, missing a contact with a specific node would not harm QuickSilver's performance as long as there are other nodes available in the same coverage range. Nonetheless, if after $2M \cdot N_c$ ($2N_c^2$ when $N_c$ is prime) hops no contact is detected although at least one channel was found free, the node randomly selects a new $s$, and continues the discovery procedure.

## 3.4 Networking

The Networking layer of QuickSilver includes two components. The first one manages intra-cluster unicast or broadcast communication with peers in the same cluster (see Figure 2). The second

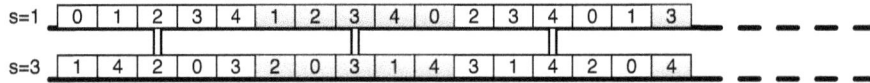

**Figure 4: Channel hopping sequence for two nodes, as long as $s$ is different the nodes will meet on every channel after $N_c^2$ hps.**

component manages the inter-cluster communication and relies on both Cluster management and Cluster Contact.

### Intra-cluster networking

Since a node's knowledge of the cluster is locally defined by its CML, intra-cluster unicast packets can only be sent to nodes that are currently in the CML of the sender node. The delivery happens using a traditional source routing approach, such as DSR. When a node generates a new message, it initiates an attempt to discover a route to the destination by flooding a RREQ message within its cluster. Relaying nodes attach their IDs to the message and propagate the broadcast. When the destination receives the message and unicasts a RREP back to the source following the reverse path through which the first RREQ reached it. The source node can then unicast the data message to the destination also using source routing. If the source node does not receive a reply in a sufficient amount of time, it assumes that the destination is not part of its cluster. To reduce the overhead of issuing a RREQ message after the creation of every new message, QuickSilver aggressively uses a route cache. To facilitate building up this cache, all messages, both control and data, are expected to contain the list of nodes traversed. Therefore, whenever a node receives any message, it can use the path contained in that message to build up routes to all of the nodes that the message has traversed. To ensure freshness, routes are timed out from the cache every after $T_{route}$ seconds.

The scope of intra-cluster broadcast messages is also limited to cluster members. The broadcast is best-effort and follows a simple flooding algorithm, with the exception that a message is accepted and rebroadcast only if it comes from a node that has been marked as a stable neighbor. To avoid congestion, each node retransmits a broadcast packet exactly once, even if the same message is received from multiple links. A log of the broadcast messages that the node most recently relayed insures this property.

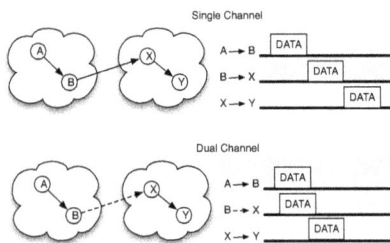

**Figure 5: Different channels for intra and inter-cluster communication mitigate the delay for multi-hop inter-cluster delivery.**

### Inter-cluster networking

When any two nodes from different clusters meet, inter-cluster communication can be initiated by setting up the fat pipe described in Section 3.2. All inter-cluster communication is multicast based on a publish-subscribe mechanism. Nodes that are interested in receiving a certain type of data (e.g., map updates, music) advertise in the c_hello messages. A node stores information about its cluster member's interests in a list similar to the CML. When two

nodes establish an inter-cluster link, they exchange their interest lists and flood them in their cluster.

In this way, each node is informed of which cluster members are gateways, and what kind of data they are trying to relay to the other cluster. When the inter-cluster link stops working, the gateway informs the cluster, so that the multicast stream is stopped or rerouted to a different gateway. It could be argued that if the links are between nodes that are not storing information but simply act as gateways for other nodes, the inter-cluster links might be underutilized while the gateways receive the data for the other cluster. The choice of using a separate channel for intra-cluster communication, as shown in Figure 5, mitigates this problem. Consider a single link between two clusters. As shown in the single channel scenario, in order to send a packet from node A to node Y, the packet has to traverse three links, and only one of them can be active at the same time to avoid collision. If instead we differentiate the channels, as in the dual channel scenario, node B can initiate transmitting the data as soon as it receives the first packet. Of course, node X might still have to wait to deliver the inter-cluster traffic to the destination, but since they are in the same cluster, this can be done at a later time. The only link that is short lived in this scenario is that between nodes B and X and QuickSilver concentrates on ensuring that that link is never underutilized.

## 4. EVALUATION

The two main components of QuickSilver are its distributed, light-weight clustering protocol and its inter-cluster multi-link connections. These two components coexist and cooperate to efficiently support intra-cluster and inter-cluster traffic. Since the multi-radio architecture and the choice of using separate channels for the two paradigms implies that there is no harmful interaction between the two components, we evaluated the components separately.

The goal of cluster management is to locally build a CML without the need for a cluster coordinator. However, the fully distributed nature of QuickSilver comes at a price: when clusters are very dynamic, a node's CML might not be accurate and the CMLs might differ from one node to another. When a node enters a cluster, the lack of an explicit join mechanism implies that for some time the cluster members are not aware of the new node and cannot communicate with it. We refer to this situation as a *false negative*. In the same way, when a node leaves a cluster for a transitory time its ID remains in the CML of the other nodes, which might still try to communicate with it, wasting resources. This is a *false positive*.

To evaluate the effectiveness of QuickSilver's cluster management, we tested it in the most unclustered environment: mobile nodes following a random way point model. Essentially, the RWP mobility model causes a very unstable clustering compared to that expected when mobility is constrained to road configurations. These evaluations show the worst-case performance of QuickSilver. However, QuickSilver can take advantage of existing traffic to improve its accuracy (i.e., when a RREQ to a node currently in the CML fails, the node can be removed even if the timeout for its entry did not expire yet). We evaluated the clustering mechanism using two metrics: the ratio between false positives and the total number of nodes in the CML per each node (FPr), and the ratio between false negatives and the total number of nodes in the cluster (FNr).

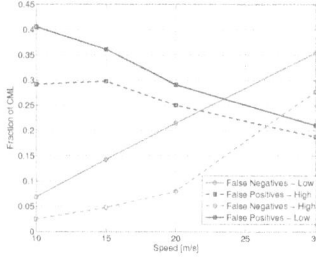

**Figure 6: Average False Positives ratio (FPr) and False Negatives ratio (FNr) for different node density and average speed.**

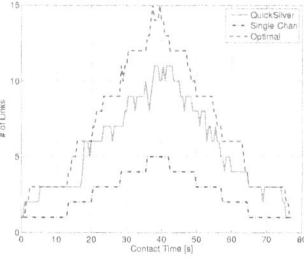

**Figure 7: Number of inter-cluster links active during a contact, compared to the optimal and the single channel solutions.**

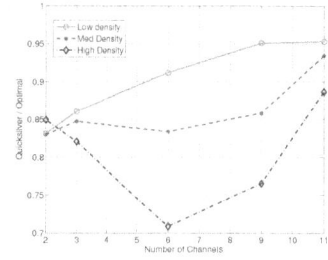

**Figure 8: Comparison between QuickSilver and the optimal solution for a cluster contact with different node density.**

The two metrics are computed per node at intervals equal to the minimum duration of a stable link, and for each run the result is an average of the metrics for all nodes over the total simulation time. A high FPr wastes resources trying to reach nodes that are no longer in the cluster. However, when real traffic is present, the lack of response to a route request can help to limit this problem. An high FNr means an increased delay after a node joins a cluster before every other node can initiate a peer-to-peer connection to it. All evaluations were performed using the ONE simulator [22] in a random mobility scenario in a $3 \times 0.2Km$ area with two levels of density (60 and 100 nodes) and different average speeds.

FPr and FNr respond differently to changes in node speed (see Figure 6): FNr increases with increased speed while FPr is lower at higher speeds. A higher node density always reduces the value of both metrics, because nodes are on average closer and partitioning is less frequent. The reason for this behavior can be explained with a simple example: consider a stable cluster with 10 nodes that suddenly splits in half. Each node will have a false positive ratio of 0.5, which is also the cluster's average FPr until the cluster stabilizes. If instead 8 of the nodes stay together and two generate a smaller cluster, the average FPr becomes $\left(2 \cdot \frac{8}{9} + 8 \cdot \frac{2}{9}\right) = 0,3\overline{5}$. At higher speeds, in the RWP mobility model, this second scenario is more likely to happen due to the high mobility. The tendency of the lower mobility scenario to have smaller clusters instead of a single large cluster and a few outliers also affects FPr. When two clusters merge in the high mobility scenario, most of the network is affected, while when the network is partitioned in small clusters, only a few nodes must update their CMLs. The RWP scenario is more representative of an urban environment, in which low speed and high density are common factors, and in this case the low FNr is desirable, while the RREQ mechanism mitigates the effect of a high FPr. In high mobility scenarios, which for VANETs generally means in highways in which the duration of a cluster membership is longer, the large initial delay caused by a high FNr has less influence.

On the second radio, nodes hop through the available orthogonal channels trying to establish a link with a node from a different cluster. For this component, QuickSilver's approach is also completely distributed, and the number of contention free links that can be established affects the total amount of data that can be transferred over an inter-cluster connection. Solutions such as [3], in which single-hop clusters are formed around a cluster coordinator, which is also the only node entitled to inter-cluster communication on a single channel, represent a lower bound for QuickSilver. Ideally, an omniscient protocol with perfect knowledge of the position of each node, can determine an optimal configuration that maximizes the number of contention free links, but at a cost in terms of computation and control overhead that cannot be sustained in dynamic

environments such as VANETs. We compared our distributed solution with these two extremes to evaluate its efficiency. It is important to note that QuickSilver is completely opportunistic and does not require any overhead. For these evaluations, we implemented the inter-cluster gateway management protocols in MATLAB.

To capture an interesting scenario, we focus on the number of contention free channels found by QuickSilver, the single-channel and the optimal approaches for a contact between two clusters traveling in opposite directions (see Figure 7). The clusters are both 1000 meters long, each composed of 30 nodes. The relative distance between a consecutive pair of nodes is modeled with a normal distribution centered around 33 meters, the speed of both clusters is $15m/s$. For this evaluation, we used 5 orthogonal channels. At $t = 0s$, only the edges of the two clusters overlap. In this configuration, there is only one pair of nodes in the reciprocal coverage range, and the three algorithms tie, with only one link possible. As the clusters keep moving and the overlapping region increases with time, the gap between the optimal solution and the single channel increases. Both solutions respect the bounds in Section 3.3. Quick-Silver's performance sits in between the two lines. QuickSilver always takes advantage of the multiple orthogonal channels, performing in some cases as well as the optimal solution, and in no situation less than 70% of it. The sub-optimal performance of Quick-Silver can be explained with two factors. First, of course, QuickSilver is a completely randomized solution, and does not have enough information to try to compute the optimal configuration. Additionally, once a link is established, QuickSilver does not drop it until the distance between the two nodes is larger than the radio coverage range. This means that even when QuickSilver reaches a configuration close to the optimal, the continued mobility soon brings the nodes to a different geographic distribution for which the current channel assignment is not optimal anymore. The optimal approach we compare against is computed for each instant, without trying to preserve the existing links.

More orthogonal channels increase the possibility of having multiple concurrent links in the same collision domain. The number of available orthogonal channels is defined by the different technologies and range from 3 for 802.11g to 16 or more for 802.11a or 802.15.4. However, this is not the only factor that influences the performance of QuickSilver. We studied the effect of channel assignment and node density by repeating the simulations for the inter-cluster contacts with low, medium and high density clusters (i.e., nodes are spaced by 65, 50 and 33 meters respectively). The ratio between the number of links established by QuickSilver and by the optimal solution is shown in Figure 8 as a function of the number of available channels. Interestingly, in the low density scenarios, a larger number of channels improves QuickSilver's performance, from 84% of the optimal with two channels up to 95%

with 9 channels. For the high density scenario, QuickSilver's performance does not keep up with the optimal scenario between 2 and 6 channels, where QuickSilver tendency to preserve existing links causes a sub-optimal configuration, in which more nodes are available but cannot establish a link because they would cause interference with the existing ones. This is a measure of how close QuickSilver performs with respect to the optimal solutions, and does not mean that the bandwidth achievable with six channels is less than that achievable with two. The results simply show that the randomized approach is less close to optimal. When the number of available channels increases again, the larger number of options helps QuickSilver to recover and approach again to the optimal solution. Although QuickSilver always uses all but one available channels for inter-cluster links, this result might suggest that better resource allocation is possible if the system is aware of nodes' density and speed. The study of the possible benefits achievable with a more complex resource management and the cost in terms of overhead to achieve it in in our research agenda.

## 5. CONCLUSION

The characteristics and requirements of applications in vehicular networks are unique and so need specifically designed protocols to unleash their potential. In this paper, we presented the design of QuickSilver, a framework that leverages clustering and data patterns typical of vehicular networks to enhance the quality of communication. In particular, QuickSilver uses a lightweight distributed clustering protocol to integrate a traditional source routing protocol for intra-cluster node-centric communication and the construction of a multi-channel link for contention-free inter-cluster data-centric communication. The limited bandwidth and short contacts typical of VANETs call for a careful resource allocation between the two communication paradigms to prevent harmful contention and wasted resources.

The benefits of using QuickSilver come from the use of a common framework for an uncoordinated set of protocols. First, both the inter-cluster and the intra-cluster benefit from the same cluster detection mechanism and avoid redundant control overhead. Second, the framework can coordinate the utilization of the available resources, preventing starvation or harmful competition. Finally, the system can react to the network dynamics and coordinate its various parts to maximize the service provided to the single node and the whole cluster. For example, when a cluster contact happens, each node running QuickSilver makes local decisions but is aware of its neighbors' behavior to establish the largest possible number of contention-free channels between the two clusters.

Although QuickSilver is able to create inter-cluster links between two clusters , more research must be done in the networking component and the resource management. We are currently working on an implementation of QuickSilver in our vehicular testbed.

[1] A. Skordylis and N. Trigoni, "Delay-bounded routing in vehicular ad-hoc networks," in *Proc. of ACM MobiHoc*, 2008.

[2] V. Namboodiri, M. Agarwal, and L. Gao, "A study on the feasibility of mobile gateways for vehicular ad-hoc networks," in *Proc. of ACM VANET*, 2004.

[3] Z. Zhang, "Routing in intermittently connected mobile ad hoc networks and delay tolerant networks: overview and challenges," *IEEE Communications Surveys Tutorials*, vol. 8, no. 1, 2006.

[4] T. Spyropoulos, K. Psounis, and C. S. Raghavendra, "Spray and wait: an efficient routing scheme for intermittently connected mobile networks," in *Proc. of ACM WDTN '05*, 2005.

[5] B. Hull, V. Bychkovsky, Y. Zhang, K. Chen, M. Goraczko, A. Miu, E. Shih, H. Balakrishnan, and S. Madden, "CarTel: a distributed mobile sensor computing system," *Proc of ACM SenSys*, 2006.

[6] K. Ibrahim and M. C. Weigle, "CASCADE: Cluster-Based Accurate Syntactic Compression of Aggregated Data in VANETs," in *Proc. of IEEE Globecom Workshops*, 2008.

[7] M. Piórkowski, N. Sarafijanovic-Djukic, and M. Grossglauser, "On clustering phenomenon in mobile partitioned networks," in *Proc. of ACM MobilityModels*, 2008.

[8] H. Dang and H. Wu, "Clustering and cluster-based routing protocol for delay-tolerant mobile networks," *IEEE Transactions on Wireless Communications*, vol. 9, no. 6, pp. 1874–1881, June 2010.

[9] J. S. Otto, F. E. Bustamante, and R. A. Berry, "Down the Block and Around the Corner The Impact of Radio Propagation on Inter-vehicle Wireless Communication," *ICDCS*, 2009.

[10] J. Burgess, B. Gallagher, D. Jensen, and B. N. Levine, "MaxProp: Routing for Vehicle-Based Disruption-Tolerant Networks," in *Proc. IEEE INFOCOM*, 2006.

[11] N. Thompson, R. Crepaldi, and R. Kravets, "Locus: a location-based data overlay for disruption-tolerant networks," in *Proc. of CHANTS*, 2010.

[12] J. Ott, E. Hyytia, P. Lassila, T. Vaegs, and J. Kangasharju, "Floating content: Information sharing in urban areas," in *Proc. of IEEE PerCom*, 2011.

[13] E. Hyytia, J. Virtamo, P. Lassila, J. Kangasharju, and J. Ott, "When does content float? Characterizing availability of anchored information in opportunistic content sharing," in *Proc. IEEE INFOCOM*, 2011.

[14] J. Wu and H. Li, "A Dominating-Set-Based Routing Scheme in Ad Hoc Wireless Networks," *Telecommunication Systems*, vol. 18, 2001.

[15] B. Das and V. Bharghavan, "Routing in ad-hoc networks using minimum connected dominating sets," in *Proc. of ICC*, vol. 1, 1997.

[16] C. Lin and M. Gerla, "Adaptive clustering for mobile wireless networks," *IEEE JSAC*, vol. 15, no. 7, pp. 1265–1275, 1997.

[17] Y. Gunter, B. Wiegel, and H. P. Grossmann, "Cluster-based Medium Access Scheme for VANETs," in *Proc. of IEEE Intelligent Transportation Systems Conference*, 2007.

[18] D. B. Johnson, D. A. Maltz, and J. Broch, "DSR: the dynamic source routing protocol for multihop wireless ad hoc networks," *Mobile Computing*, vol. 353, 2001.

[19] C. Perkins and E. Belding-Royer, "Ad hoc On-Demand Distance Vector (AODV) Routing," in *Proc. of IEEE WMCSA*, 1999.

[20] A. Lindgren, A. Doria, and O. Schelén, "Probabilistic routing in intermittently connected networks," *ACM SIGMOBILE Mobile Computing and Communications Review*, vol. 7, 2003.

[21] P. Bahl, R. Chandra, and J. Dunagan, "SSCH: slotted seeded channel hopping for capacity improvement in IEEE 802.11 ad-hoc wireless networks," in *Proc. of MobiCom*, 2004.

[22] A. Keränen, "The ONE Simulator for DTN Protocol Evaluation," in *Proc. ICST SimuTools*, 2009.

# Demo of a Collaborative Music Sharing System

Zhaofei Chen, Emre A. Yavuz, Gunnar Karlsson
Laboratory for Communication Networks
Linnaeus Center ACCESS
KTH, Stockholm, Sweden
{zhaofei, emreya, gk}@kth.se

## ABSTRACT

We demonstrate a wireless real-time music-sharing application that lets users play music directly from their mobiles through a jukebox. We have designed and implemented the application by using a previously developed content-centric opportunistic networking middleware. The jukebox plays the music file that is first in its playlist by streaming it in real-time from the publishing user device. All users can observe the collaboratively formed playlist on their mobiles in real-time. The application shows the usefulness of our middleware and demonstrates a new form of situated applications. The application handles churn and garbage collection after departed users.

## Categories and Subject Descriptors

C.2.1 [**Network Architecture and Design** ]: Wireless communication

## General Terms

Design

## Keywords

Opportunistic Networking, Peer-to-peer, Publish/subscribe, DTN

## 1. INTRODUCTION

The fast spreading of smart mobile devices has changed the way people create and share multimedia contents. In order to explore this, we have designed a collaborative music sharing system over an opportunistic ad hoc network of mobile devices. The users of the mobile devices in the system can share music through a central jukebox in real-time. The session management is carried out by means of our middleware for opportunistic networking while the streaming is from each mobile through the jukebox; only a few bytes of data, not the entire music files, reside in the jukebox. It is a publish-subscribe system in which the jukebox acts as the sole subscriber for contents published by the mobiles and the mobiles act as the subscribers for the playlist published by the jukebox.

The main motivation for this system is to enable users to share contents, in this case music files stored in the mobile devices, in real-time (the system could be extended to work for video, images and more). In contrast to other wireless ad hoc content sharing systems, here the concept of sharing does not aim at exchanging the music files between neighboring nodes but rather to stream them to

a central unit one at a time so that others can enjoy the music as long as they remain in proximity of the jukebox. Typical venues to setup such a system may be public places where people gather for entertainment, such as bars and clubs, and in private homes. The system has been developed for two different reasons. First, we would like to demonstrate the versatility of our opportunistic, content-centric middleware for use in services that only are and should be locally available to those situated at a certain location or within a given premise (as opposed to the internet idea of making everything accessible by everyone everywhere). The content-centric concept is natural since identities of devices are irrelevant, and what matters are only the contents sought. The second motivation is to test this particular system to see whether sharing of public address systems could be of interest, how it should be done, and how well it may work.

Section 2 of the paper gives an overview of the architecture, with the implementation following in Section 3. The demonstration scenario is presented in Section 4 and the conclusion in Section 5.

## 2. SYSTEM ARCHITECTURE

The overview of our music sharing system is shown in Fig. 1. The system has two main actors: A device that is connected to a sound system, on which the jukebox application resides (a stationary computer or docked mobile device), and mobile devices that are within direct communication range by means of some short-range radio, like WiFi. The circle in Fig. 1 indicates the communication range between the jukebox and the mobile devices. Every time a device is in range, it can establish a direct link to the jukebox over which the user may play music that he or she wishes to share with others. Multi-hopping is not provided in order to limit the spatial scope of the service, to avoid poor quality streaming, and to avoid the complexity of routing in environments with mobility and churn in the user population (bars, cafés and clubs). A playlist is created from the user contributions and it is published by the jukebox.

The system modules in both the jukebox and the mobile devices are illustrated in Fig. 2. The *jukebox application* module provides users/admins a UI to control the jukebox manually when needed (e.g., to delete songs). The *RTSP streaming client* module enables the streaming service in cooperation with the *RTSP streaming server* module in the mobile devices. The *playlist handler* acts as the database of the Podnet [5] service running on the jukebox and it manages the playlist.

The mobile devices contain three modules. The *mobile application* module is basically an Android application that provides a UI to users for publishing music files and monitoring the playlist. Each user device has an *RTSP streaming server* module since shared music files are streamed in real-time. The *Podnet middleware* enables the content exchange between the user devices and the jukebox. It

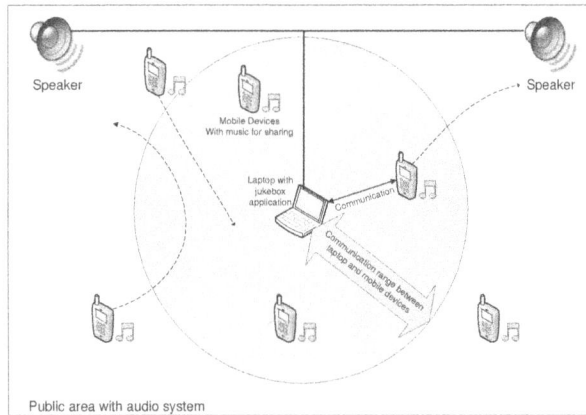

**Figure 1: Overview of the collaborative music sharing system.**

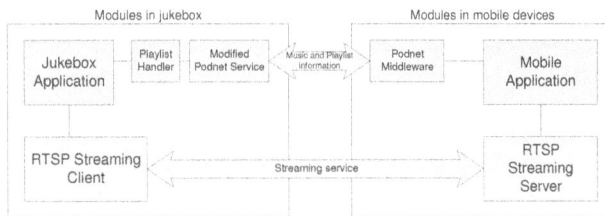

**Figure 2: Modules of the collaborative music sharing system.**

provides a publish/subscribe interface to applications that bind to it. Contents that users share are organized into *feeds* according to the *Atom* data structure. Our system has two feed, *music* and *playlist*, for all mobiles to publish music suggestions and to synchronize their playlists with the jukebox. The jukebox is subscribed to the *music* feed to receive streaming requests from the users and it publishes the playlist to the *playlist* feed whenever updated.

## 3. IMPLEMENTATION

The designed architecture has been implemented in Java and tested on HTC Hero mobile devices and a stationary computer running on the Android and Ubuntu OS platforms, respectively. The implementation is based on IEEE 802.11 in ad-hoc mode which is not supported by the Android Java libraries. Both the driver and the hardware interface on the HTC Hero device can be run in ad-hoc mode, but it must then be run in privileged user mode (i.e. rooted mode). On the stationary computer, the jukebox application module uses Jajuk [1] which is a full-featured open source jukebox application written in Java. Various UI components and perspective designs are used for a more convenient user handling. Jajuk works with an open-source audio/movie player called MPlayer [3] in Ubuntu that is called every time the jukebox needs to play a music file.

MPlayer does not support RTSP streaming and hence it needs to be recompiled with an appropriate media library that supports real-time streaming. We recompiled the source code of the MPlayer application with the *Live555 Streaming Media* libraries, which are a set of source-code libraries for standards-based RTP/RTCP/RTSP/SIP multimedia streaming [2]. We implemented a RTSP server and installed it on the mobile devices to provide RTSP streaming service.

More information about the design and implementation of the system and the evaluation of it is available in [4].

## 4. DEMONSTRATION

We demonstrate a scenario where we have a stationary computer running the jukebox application and users who are equipped with HTC Hero mobiles that contain different sets of music. We let users play any of the available music files on the mobile devices given to them by tapping the *Add Music* button implemented as a *Menu* function, selecting a file, and publishing it. The title of the selected song is then displayed both on the *Published* tab of the corresponding device's main screen and on the *Queue View* of the jukebox running on the computer. The *Queue View* allows users to see the playlist that displays all music suggestions published by the users in their order of arrival. The *Playlist* tab on the user devices displays the playlist in synchronization with the jukebox so that users can monitor the queueing status of their own as well as all suggestions that have been published. A *Playing* tag marks the currently playing music file in the playlist.

We present the following additional scenarios to show that our system has a robust design. When the jukebox application is turned off intentionally (or unexpectedly due to a system crash or power failure in real life), the streaming stops immediately. But the system status is restored when it is turned on again and the jukebox starts streaming from the beginning of the music file that was playing when the system was turned off. If a similar situation is experienced by a mobile device from which the jukebox streams (for instance the user could have left the venue), the jukebox automatically skips the song and plays song the next in the list.

## 5. CONCLUSION

We demonstrate a collective music sharing system that allows users to submit music suggestions to a jukebox within radio range. The associated music files, residing on users' mobile devices, are played in order of the compiled playlist via real time streaming upon request from the jukebox. Our system has two components: a jukebox application that runs on a stationary computer or a docked mobile device connected to a sound system, and a mobile application that runs on Android smart phones. Both applications are designed and implemented to utilize the Podnet content-centric opportunistic middleware for session management. A playlist is distributed to all participating devices to keep the users updated on coming songs. The system is designed to handle churn in the user population.

## 6. REFERENCES

[1] Jajuk Advanced Jukebox - Official Site, 2011.
    http://jajuk.info/index.php/Main_Page.
[2] LIVE555 - Streaming Media, 2011.
    http://www.live555.com/liveMedia/.
[3] Mplayer - The Movie Player, 2011.
    http://www.mplayerhq.hu/design7/info.html.
[4] Z. Chen. Opportunistic crowdsourcing for collective music sharing. Master's thesis, KTH–Royal Institute of Technology, School of Electrical Engineering, August 2011.
[5] Ó. Helgason, E. A. Yavuz, S. Kouyoumdjieva, L. Pajevic, and G. Karlsson. A mobile peer-to-peer system for opportunistic content-centric networking. In *Proc. ACM SIGCOMM MobiHeld workshop*, 2010.

# PePiT: Opportunistic Dissemination of Large Contents on Android Mobile Devices

Matteo Sammarco\*, Nadjet Belblidia\*, Yoann Lopez°, Marcelo Dias de Amorim\*,
Luis Henrique M. K. Costa†, and Jérémie Leguay°

\* CNRS and UPMC Sorbonne Universités    ° Thales Communications & Security    † Universidade Federal do Rio de Janeiro
Paris, France                            Colombes, France                      Rio de Janeiro, Brasil
firstname.lastname@lip6.fr               firstname.lastname@thalesgroup.com     luish@gta.ufrj.br

## Categories and Subject Descriptors

C.2.4 [**Computer-Communication Network**]: Distributed Systems—*Distributed applications*

## 1. CONTEXT AND MOTIVATION

Enabling content sharing among mobile users is a promising application for opportunistic networks. Clearly, collocated people are likely to share mutual interests. In this context, disseminating contents through opportunistic communications could be more efficient than passing through central servers.

We implemented PACS, a popularity-based strategy to select pieces of contents to be exchanged between collocated devices solely based on local information [1]. Through their successive contacts, devices keep track of the dissemination level of the pieces throughout the network and use this information to transfer less prevalent pieces first.

In this paper, we present PePiT, an Android application based on PACS. PePiT enables the dissemination of pictures between collocated Android devices in an ad hoc mode. We show both the architecture and the deployment requirement of PePiT on Android devices. We also briefly describe the demonstration scenario.

## 2. PACS: PREVALENCE-AWARE CONTENT SPREADING

The goals of PACS are to achieve fast content dissemination while keeping the overhead low and making better use of contact opportunities [1].

Let $\mathbf{N} = \{n_0, n_1, \ldots, n_{N-1}\}$ be the set of $N$ nodes in the network. Nodes are mobile, but we do not assume any a priori knowledge of mobility patterns. To illustrate the workings of the algorithm, we assume that all nodes in the network are interested in the unique content initially available at a single node called the *data source*. To generalize to any number of data sources and contents, we simply apply the algorithm to a randomly selected content.

The data source chops the content into $K$ pieces of equal size. Pieces are sequentially identified as $c_0, c_1, \ldots, c_{K-1}$. Nodes use their contact opportunities to get pieces (i.e., we assume that there is no infrastructure to help the dissemination process). Nodes can get pieces from the data source and from any other node in the network having it. Each

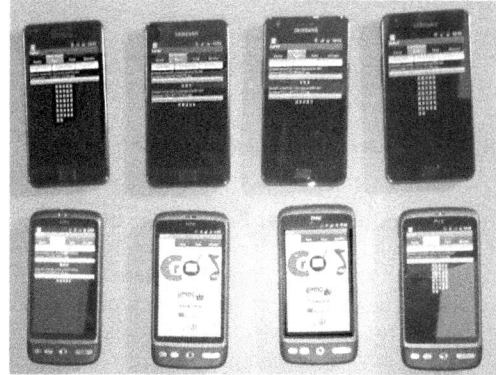

Figure 1: Mobile devices running PePiT.

node $n_i$ locally stores an *availability bitmap vector* $\mathbf{a}_{n_i} = \{a_0, \ldots, a_{K-1}\}$, where $a_k = 1$ if the node has piece $c_k$, and $a_k = 0$ otherwise. In addition to the availability vector, node $n_i$ also keeps a prevalence vector $\mathbf{p}_{n_i} = \{p_0, p_1, \ldots, p_{K-1}\}$ that gives a local view of the prevalent pieces in the network. Initially, all nodes have an empty prevalence vector.

When nodes $n_i$ and $n_j$ meet, they exchange their availability vectors $\mathbf{a}_{n_i}$ and $\mathbf{a}_{n_j}$. Node $n_i$ (resp. $n_j$) computes $\mathbf{a}_{n_i} \wedge (\neg \mathbf{a}_{n_j})$ (resp. $\mathbf{a}_{n_j} \wedge (\neg \mathbf{a}_{n_i})$), which gives the candidate pieces to be transferred ($\wedge$ stands for the "AND" operator and $\neg$ for "NOT"). They also update their prevalence vectors respectively as: $\mathbf{p}_{n_i} \leftarrow \mathbf{p}_{n_i} + \mathbf{a}_{n_j}$, and $\mathbf{p}_{n_j} \leftarrow \mathbf{p}_{n_j} + \mathbf{a}_{n_i}$.

Among the candidate pieces to be transferred, nodes select the one with the lowest prevalence. In case of tie, a piece is chosen in a uniformly distributed random way. Let $c_{i \rightarrow j}$ be the piece sent by $n_i$ to $n_j$ and $c_{j \rightarrow i}$ be the piece sent by $n_j$ to $n_i$. After one round of exchanges, nodes update their availability vectors as: $\mathbf{a}_{n_i} \leftarrow \mathbf{a}_{n_i} \vee \mathbf{i}_{c_{j \rightarrow i}}$, and $\mathbf{a}_{n_j} \leftarrow \mathbf{a}_{n_j} \vee \mathbf{i}_{c_{i \rightarrow j}}$, where $\mathbf{i}_{c_{i \rightarrow j}}$ and $\mathbf{i}_{c_{j \rightarrow i}}$ are $K$ element vectors with all positions set to 0 except the position relative to the piece just received, which is set to 1 ($\vee$ is the "OR" operator).

## 3. PEPIT: DESIGN AND IMPLEMENTATION

In this section, we present all the steps to bring PACS from theory to practice with the PePiT Android application.

### 3.1 Architecture

A mobile application, called PePiT, has been developed for Android operative systems using JAVA. It is mainly com-

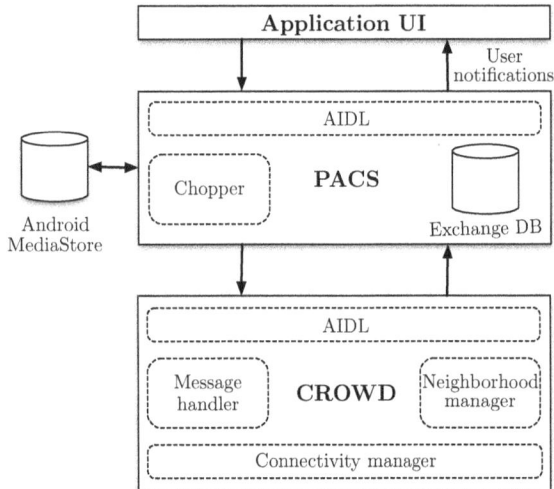

**Figure 2: Extensible architecture of PePiT.**

**Figure 3: PePiT. From left to right: exchanges in progress, received pieces from different peers historical, preview of the received picture.**

posed by two modules, called PACS and CROWD[1], implemented as Android remote services (Fig. 2). In this way, taking advantage of the Android Interface Definition Language (AIDL), they can be called by external applications making the system extensible and able to exchange photos, videos, files, or any other content. We decoupled the two modules so that the lower level functionalities of CROWD could be used by differnt higher modules or protocols.

**Application User Interface.** The application user interface provides the possibility to exchange a photo either by picking it out from the gallery or by directly shooting it with the phone camera.

**PACS Component.** The *chopper* sub-module cuts the photo into $K$ pieces filling the *availability bitmap vector* and creating the *prevalence vector*. The PACS service instantiates the PACS protocol and runs in the background. As such, it selects and sends content pieces using the CROWD service. Also, it tracks every received piece into an internal exchange database.

**CROWD Component.** CROWD service is composed of three sub-modules. The *connectivity manager* sub-module creates (or connects to) an ad hoc network, and offers an interface for sending and receiving packets through UDP or TCP, in unicast or broadcast. The *neighborhood manager* broadcasts UDP beacons every $T$ seconds in order to announce its presence to the neighbors. It also keeps state of the neighboring devices. Finally, the *message handler* sub-module creates and serializes outgoing CROWD messages and parses incoming ones.

## 3.2 Deployment on Android mobile device

PePiT runs on mobile phones equipped with Android system with a minimum API level equal to 8 (Android Froyo, version 2.2.x). It has been successfully installed on a virtual machine (VM) running android-x86 [2].

Because the Android system does not provide an API to manage IEEE 802.11 ad hoc communications, it is foremost required to follow a procedure, called *rooting*, in order

to gain administrative access rights on the phones. Taking advantage of the Android NDK (Native Development Kit) tools, we also compiled *Linux wireless tools* for ARM and wrapped them into PePiT. In this way it is possible to connect the smartphone to an ad hoc network just like in a Linux environment. As a future work, we could use a method like WiFi-Opp [3] or the new Wi-Fi Direct [4] to make the application runnable also on non rooted stock smartphones.

## 4. DEMONSTRATION

For the demonstration, we provide 4 HTC Desire and 8 Samsung Galaxy-S-II, equipped with Android 2.3.3, as well as a laptop with a virtual machine running android-x86 with Android 2.2.1. While the android-x86 screen is projected, the phones can be distributed to some people in the public.

Once PePiT is started, the mobile phones connect to the *CrowdAdHoc* ad hoc network and auto-configure an IPv4 address. They are then ready to exchange pictures to other phones in their vicinity.

For all exchanged contents, a progress bar is displayed (Fig. 3) with the detail of the missing and received pieces. When the download is complete, the picture is stored in the phone gallery and a preview is available by clicking on the item.

## Acknowledgment

This work is partially supported by the ANR Crowd project under contract ANR-08-VERS-006 and CNPq and FAPERJ Brazilian agencies.

## 5. REFERENCES

[1] N. Belblidia, M. Dias de Amorim, L. H. M. K. Costa, J. Leguay, and V. Conan, "PACS: Chopping and shuffling large contents for faster opportunistic dissemination," in *WONS*, Bardonecchia, Italy, 2011.

[2] [Online]. Available: http://www.android-x86.org

[3] S. Trifunovic, B. Distl, D. Schatzmann, and F. Legendre, "Wifi-opp: ad-hoc-less opportunistic networking," in *ACM CHANTS*, Las Vegas, Nevada, USA, 2011, pp. 37–42.

[4] [Online]. Available: http://www.wi-fi.org/Wi-Fi_Direct.php

---

[1]CROWD is the name of the project that supports this work (http://anr-crowd.lip6.fr).

# Mobile Video Broadcasting System with Multiple Mobile Devices for Improving Availability

Takeshi Horiuchi, Takuya Omizo, and Katsuyoshi Iida
Tokyo Institute of Technology
{horiuchi, mizo}@netsys.ss.titech.ac.jp, iida@gsic.titech.ac.jp

## ABSTRACT

We propose a mobile video broadcasting system, which uses multiple mobile devices, to improve the availability. Our key idea is to notify the spare device information to allow sudden shutdowns of the active device. After implemented our system, we develop a demo system to show the ability of our system. Our idea will be useful in broad applications in mobile opportunistic networks.

## Categories and Subject Descriptors

C.2.1 [**Computer-Communication Networks**]: NetworkArchitecture and Design— *Wireless Communication*
; C.2.2 [**Computer-Communication Networks**]: Network Protocols— *Application*

## General Terms

Design, Implementation

## Keywords

Mobile video broadcasting, Availability, SIP, Advanced notification method

## 1. INTRODUCTION

Recently, mobile video broadcasting services are widely spreading, i.e., camera and communication interface equipped mobile devices, such as smartphones, can easily broadcast video images through video broadcasting services such as USTREAM. The most significant issue related with mobile video broadcasting services is to provide their availability to avoid sudden disconnection of video images. This is because the limitation of battery capacity and unreliable communication media can degrade the availability of mobile video broadcasting services. If we cannot provide sufficient level of the availability, the audience may not tolerate the quality-of-service. Another issue is to provide a lightweight system because users of mobile video broadcasting services generally cannot prepare regular video equipments like TV broadcasting stations. In summary, a lightweight system to improve the availability is necessary for mobile broadcasting services.

To achieve such a requirement, we take an approach related to mobile opportunistic networks i.e., we prepare multiple mobile devices to switch over when an active mobile

broadcasting device[1] shuts off. This can reduce the length of unavailable time. To switch the broadcasting mobile device, a session migration technology may be useful to continue the service. In this paper, we propose a novel mobile broadcasting system automatically to switch the spare device to minimize the unavailable time. Our broadcasting system uses a session migration technology with enhancements of session initiation protocol (SIP) to achieve the automatic switching. In the reminder of this paper, we describe the detail of our proposed system, its implementation as well as demo description.

## 2. PROPOSED MOBILE VIDEO BROADCASTING SYSTEM

As we described in the Introduction, to improve the availability of mobile video broadcasting services, we use multiple mobile devices, i.e., the active and spare devices. Automatically to switch over the mobile devices after a sudden disconnection of the active device occurs, we employ a session migration technology. In this section, we describe an overview of our proposed mobile video broadcasting system.

We first give our system description such as mobile devices, and assumed communication environments including video distribution networks. (See Fig. 1.) As we described, we prepare the active and spare devices to improve the availability of the broadcasting service. To deliver video traffic, we assume a video distribution network such as content delivery networks (CDNs). More specifically, video traffic generated by mobile device is delivered to a video distribution server as a relay node, which relays the video traffic to the clients through the video distribution network.

To carry out the automatic switching, we have to satisfy two requirements; a disconnection detection mechanism and a switching mechanism to one of the spare devices. For the former requirement, we introduce a video packet monitoring method at the video distribution server. To satisfy the later requirement, we propose an advance notification method for the spare devices to the video distribution server before a disconnection actually happening. We then describe those two methods in detail.

The video packet monitoring method is used to detect sudden disconnection of the active device. More specifically, we apply a threshold period in terms of disconnection period. If the observed disconnection period at the video distribution server is longer than the threshold, the video distribution server detects the disconnection of the active device.

---

[1]Hereafeter, we describe "active device" to represent active mobile broadcasting device in short.

*MobiOpp'12*, March 15–16, 2012, Zürich, Switzerland.
ACM 978-1-4503-1208-0/12/03.

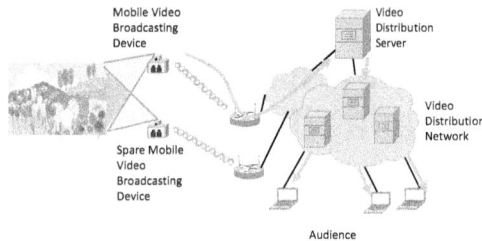

Figure 1: Proposed mobile broadcasting system

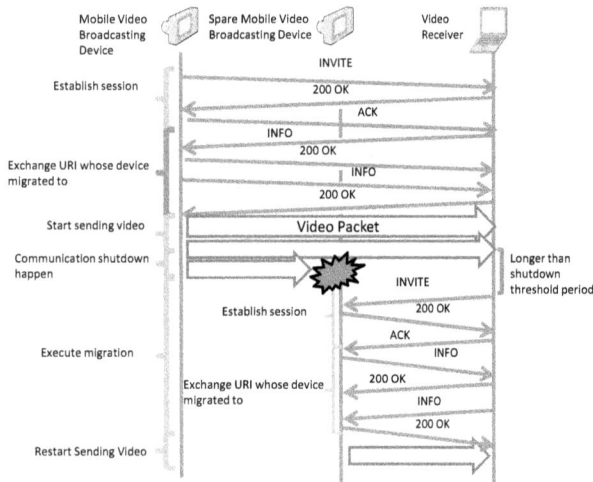

Figure 2: SIP sequence in proposed system

We then go into the next method, i.e., the advanced notification method. The difficulty to establish the switching mechanism is to identify the spare devices. In the standard session migration mechanism, the migration signal is sent from the current active device[1]. However, we would like to support the sudden disconnection of the active device, so that we need to notify the address information of the spare devices in advance. Therefore, we propose the advanced notification method to inform the addresses of the spare devices to the video distribution server to prepare the sudden disconnections of the active device. After detection of the active device disconnection by the monitoring method, the video distribution server will try to switch the spare device, in which the address of it is notified in advance.

## 3. IMPLEMENTATION

In this section, we describe the implementation of our proposed system.

As we described, the advanced notification method is necessary to prepare sudden disconnections of the active device. To implement the advanced notification method, we employ INFO method, which can be used to exchange enhanced information between clients[2], to notify the URI to reach the spare device to the video receiving software. Another protocol enhancement we require is to switch over the spare device after detection of the active device disconnections. These protocol enhancements are illustrated in Fig. 2.

We then go into the implementation of softwares on the mobile video broadcasting device such as the active and spare devices. The mobile video broadcasting device should accommodate a SIP client and a video sending application. For the video receiving software, we use a SIP client as well as a video player application with some interactions between

Table 1: Used operating systems and softwares

| Mobilie Video Broadcasting Devices | |
|---|---|
| OS | Android 2.2 |
| SIP Library | MjSIP |
| Video Streamer | Modifying class in sipdroid |
| Video Receiver | |
| OS | Windows XP SP3 |
| SIP Library | osip & eXosip |
| Player Library | FFMpeg & SDL |

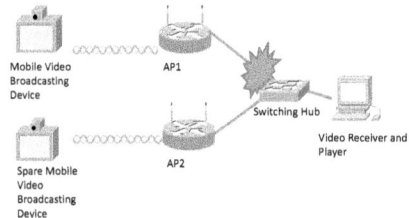

Figure 3: Demo illustration

each other. Our used operating systems and softwares are described in Table 1.

## 4. DEMO DESCRIPTION

In this section, we explain our demo using the implementation described in the former section.

The topology of the demo is illustrated in Fig. 3. We provide two mobile video broadcasting devices including the active and spare devices, each of which is connected to a WiFi access point (AP) to connect to the video receiving terminal. To represent the sudden disconnection of the active device, we remove the network cable on the AP1. After removing the cable, our disconnection detection mechanism and switching mechanism will be performed to switch over the spare device.

From the viewpoint of the video receiving terminal, it experiences some disconnection periods, but it automatically recovers the video playout received from the spare device. Note that our demo system allows a disconnection of a cable on AP2 to switch over to the upper device.

## 5. CONCLUSION

We have proposed a mobile video broadcasting system to improve the availability. Our key idea is to notify the spare device information to allow sudden shutdowns of the active device. After implemented our system, we have developed a demo system to show the ability of our system. Our approach will be useful in broad applications in mobile opportunistic networks having intermittent connection.

## Acknowledgments

This work was supported in part by Grant-in-Aid for Young Scientists (A) (1) (21680006) of the Ministry of Education, Culture, Sports, Science and Technology, Japan, and in part by NEC corporation, in Japan.

## 6. REFERENCES

[1] R. Sparks, "The Session Initiation Protocol (SIP) Refer Method," *IETF RFC3515*, 2003.
[2] N. Imai, M. Isomura, H. Horiuchi, and S. Obana, "Service Initiation and Migration for Real-time Communication Services in the Ubiquitous Network Environment," *IPSJ Journal*, 45(12), pp. 2630–2641, Dec. 2004.

# Opportunistic Networks and Cognitive Management Systems in the Service of the Future Internet: Indicative Scenarios of Coverage and Capacity Extension

Panagiotis Demestichas, Nikos Koutsouris, Vera Stavroulaki
University of Piraeus
80 Karaoli & Dimitriou Street
185 34 Piraeus, GREECE
Tel: + 30 210 414 2758

{pdemest,nkouts,veras}@unipi.gr

## ABSTRACT

In this demonstration proposal we showcase a solution that addresses the main requirements posed by the Future Internet era, such as the expanded use of wireless access and the need for increased efficiency in resource provisioning. We exploit the great potentialities of opportunistic networks, which can be seen as operator-governed, temporary and probably infrastructure-less extensions of the infrastructure-based network, and we combine them with cognitive systems, both for managing the opportunistic networks and for coordinating with the infrastructure, through the use of control channels. This prototype implementation can show the merits of opportunistic networks in terms of enhanced service provision capabilities, higher resource utilization, lower transmission powers, and "green" network operation.

## Categories and Subject Descriptors

C.2.1 [**Network Architecture and Design**]: Wireless communication, C.2.3 [**Network Operations**]: Network management.

## General Terms

Algorithms, Management, Design, Experimentation, Standardization, Verification.

## Keywords

Opportunistic networks, cognitive management, coverage extension, capacity extension

## 1. INTRODUCTION

This document is a submission for a demonstration stemming from the OneFIT (Opportunistic networks and Cognitive Management Systems for Efficient Application Provision in the Future Internet) project [1]. The demonstration will showcase architectures that enable the cost efficient provision of a wide range of diversified applications at the appropriate quality levels, based on the amalgamation and cooperation of network infrastructures and opportunistic networks. Further information on

the demonstration and the OneFIT project, including the planned validation activities can be found in [1]. The project capitalizes on the results of [2]. The demonstration also covers aspects developed in [3].

## 2. SYSTEM DESCRIPTION IN BRIEF

The OneFIT prototype system encompasses mainly the following technologies: opportunistic networks, cognitive systems for managing the opportunistic network and the infrastructure and Control Channels for the Cooperation of the Cognitive Management Systems (C4MS).

## 2.1 Opportunistic Networks

Opportunistic Networks (ONs) are operator-governed, temporary, coordinated extensions of the infrastructure. An ON is created dynamically, whenever and wherever it is needed in order to deliver multimedia flows to mobile users, in a cost-efficient manner, respecting at the same time at least a certain Quality of Experience and a minimum set of Quality of Service levels. In any case the operator designates the resources (e.g., spectrum, transmission powers, etc.) that can be used, and governs through the provision of policies, information and knowledge. ONs can comprise network elements of the infrastructure, and terminals/devices potentially organized in an ad-hoc network mode.

## 2.2 Cognitive systems

The management of the opportunistic networks and the necessary coordination with the infrastructure is efficiently performed by cognitive management systems. They provide the means for the suitability determination, creation, modification and release of opportunistic networks. There are two types of systems in the presented architecture, the "Cognitive systems for Managing the Opportunistic Network" (CMONs) and the "Cognitive management Systems for Coordinating the Infrastructure" (CSCIs).

Both have a similar high-level structure, comprising context, profiles, policies, negotiation, optimization, decision-making and learning capabilities. However, they play primary roles in different phases. The CSCI have the primary role in feasibility determination (even though it will cooperate with the CMONs). The CMONs, on the other hand, have a primary role in the creation and maintenance of the opportunistic networks, within the framework set by the CSCI. A fundamental idea of the OneFIT concept is to provide the means to facilitate close cooperation between the infrastructure and the opportunistic networks. Such collaboration is essential for ensuring viability, deployment and value creation for all the stakeholders.

## 2.3 Control channels

C4MS is an evolution of the Cognitive Pilot Channel (CPC) concept. CPC was mainly developed for the cooperation of the infrastructure with devices in a flexible spectrum management concept. C4MS brings the content of this cooperation several steps further, by including general context, profiles, policies, knowledge and decisions. C4MS relies on application level interactions between the cognitive management systems and its implementation can involve direct, inter-device communication. Relevant specification and standardization activities are conducted in [3].

## 3. NOVEL CHARACTERISTICS

The OneFIT prototype system includes several novel features related to the following topics:

☐ The cost-efficient provision of main applications envisaged in the Future Internet context;

☐ The coordinated operation between heterogeneous infrastructures and the opportunistic network;

☐ Cognitive management systems (i.e., encompassing self-management and knowledge) and control channels for the optimal operation of the opportunistic network and the infrastructure.

The efficiency of the system is proven based on facts such as: (i) the higher utilization of resources, and therefore, the achievement of higher capacity levels, without the need for new investment in the infrastructure; this means lower CAPEX; (ii) the use of lower transmission powers by the infrastructure, and therefore, the lower energy consumption in the infrastructure; this means lower OPEX; (iii) the high "green" footprint of the application delivery model. The facts above are also validation criteria for OneFIT.

## 4. DEMONSTRATION SCENARIOS

Through the proposed demonstration various use cases will be illustrated, according to the following scenarios:

☐ Application delivery by expanding the coverage of the infrastructure through opportunistic networks.

☐ Application delivery by resolving cases of congested access to the infrastructure through opportunistic networks (compared to the mere reconfiguration or long term re-planning of the infrastructure).

Demonstrations will include software prototypes of the cognitive management systems and the control channels. As already presented cognitive management systems involve context, profiles, policies, negotiation, optimization, decision-making and learning capabilities. Control channels provide the means for the exchange of information and knowledge on these capabilities.

## 5. IMPLEMENTATION ASPECTS

The OneFIT prototype is implemented as a multi agent system and it is mainly based on Java and the JADE [4] middleware platform. The demo is distributed among several laptops that play

the role of network elements or terminals. There is also management software installed in commercial mobile devices (PDAs). All the elements that are involved in the scenarios, emulated or available as real hardware, are parts of the same platform and their interactions, decisions and actions are visualized through the graphical user interface shown in Figure 1.

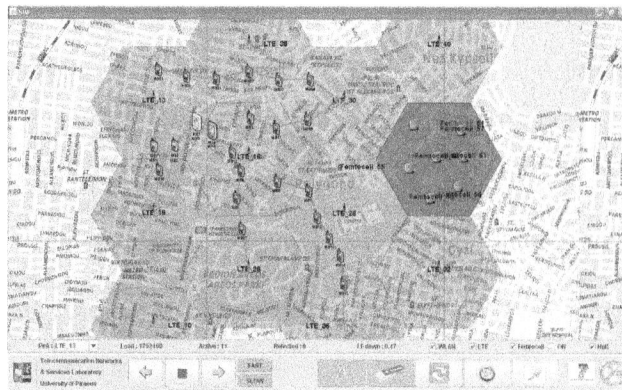

**Figure 1. OneFIT platform implementation.**

## 6. CONCLUSIONS

This document is a proposal for a demonstration stemming from the OneFIT project. The document included a short description of the system, a summary of its novel characteristics, and the functions and features that will be demonstrated.

## 7. ACKNOWLEDGMENTS

This work is performed in the project OneFIT which is partially funded by the Community's Seventh Framework programme. This paper reflects only the authors' views and the Community is not liable for any use that may be made of the information contained therein. The contributions of colleagues from the OneFIT consortium are hereby acknowledged.

## 8. REFERENCES

[1] OneFIT (Opportunistic networks and Cognitive Management Systems for Efficient Application Provision in the Future Internet) project, www.ict-onefit.eu, 7th Framework Programme (FP7) for Research and Technological Development, on Information and Communication Technologies, of the European Commission.

[2] E3 (End-to-End Efficiency) project, www.ict-e3.eu, 7th Framework Programme (FP7) for Research and Technological Development, on Information and Communication Technologies, of the European Commission.

[3] European Telecommunication Standardization Institute (ETSI), Technical Committee (TC) "Reconfigurable Radio Systems" (RRS), Web site http://www.etsi.org/website/technologies/RRS.aspx.

[4] Java Agent DEvelopment Framework (JADE), Web site http://jade.cselt.it.

# Illinois Vehicular Project, Live Data Sampling and Opportunistic Internet Connectivity

Riccardo Crepaldi, Ryan Beavers, Braden Ehrat, Robin Kravets
University of Illinois at Urbana-Champaign
{rcrepal2,beavers1, ehrat1, rhk}@illinois.edu

## Categories and Subject Descriptors

C.2.1 [**Computer-Communication Networks**]: Network Architecture and Design—*Distributed networks,Store and forward networks*; C.2.4 [**Computer-Communication Networks**]: Distributed Systems

## Keywords

VANETs, Experimental Testbed, Opportunistic Connectivity, Ad-Hoc

Embedded sensors in mobile devices such as cars and smart phones present new opportunities to collect data and explore a new environment. Vehicular Networks are highly mobile and widely spread, and the broad deployment of embedded sensors will lead to the establishment of large participatory sensing systems and enable the generation of large amounts of data. A major challenge is efficiently collecting, storing and sharing all this data.

Vehicular networks present several bottlenecks that must be considered. Data could be kept locally using the network as a distributed storage system. However, the high mobility and frequent disconnection could cause a continuos migration of data, and possibly its loss. Replication is a possible solution to this problem, using ad-hoc connections between vehicles. Additionally, capacity is also a challenge. Using a mobile connection such as 3G or WiMax, the information can be uploaded and retrieved from a central storage unit. However, those networks are already pushed to the limit to serve existing mobile Internet access. The increasing number of devices generating data and the rates at which this data is generated will quickly overwhelm the infrastructure. On the other hand, energy efficiency might not be a primary concern for vehicles, given that when a vehicle is on, it generates enough energy for full-power radio operation. Nevertheless, a vehicle can be parked for several days, and thus the power for communication would drain the battery. Thus, even in vehicular networks, energy efficient protocols can and sometimes must be used.

Efforts to tackle these challenges led to the design of systems such as Locus [4]. However, it is still not fully understood how will these heavy sensing tasks, the peer to peer communication, and the energy efficiency interact. To an-

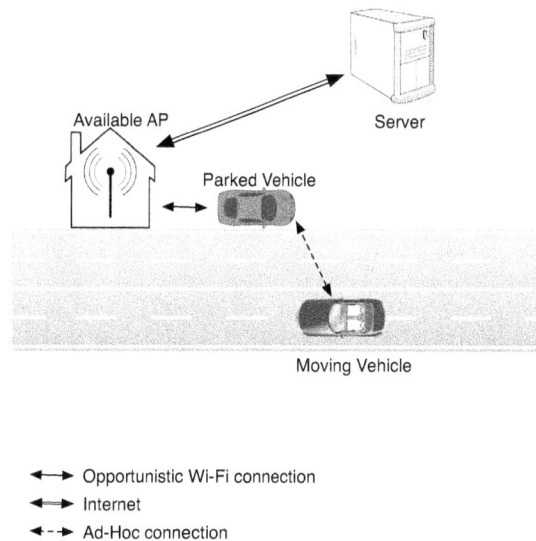

Figure 1: Demo architecture

swer this question, among others, we designed and deployed the Illinois Vehicular Testbed (IVT). We instrumented a number of cars with wireless radios (802.11n/g) for car-to-car and car-to-server communication, Bluetooth, GPS receiver, OBD-II receiver to collect on-board diagnostics from the vehicle.

Our platform was designed around low powered Geode processors with an emphasis on efficiency. In lieu of traditional hard disk drives, our systems utilize flash memory for local storage and require limited read/write accesses to reduce slow downs. With a memory ceiling of 256MB, we have an environment that requires an acute focus on memory usage. The IVT is running a customized Linux kernel to reduce its memory footprint and support its modular design. All of these restrictions allow us to isolate the effects of the data transfer and study it effectively. On each unit, the 802.11n radio is dedicated to Internet connection by attaching to existing access points, while the 802.11g is used for car-to-car communication. The challenge for the latter is that, while moving vehicles have no energy constraints and can have their radio on the whole time, and therefore is easy to establish a car-to-car connection, those that are parked must use a sleep schedule to prevent excessive battery depletion, thus making these vehicles harder to discover. An appropriate duty-cycling-based discovery mechanism must

**Illinois Vehicular Project**

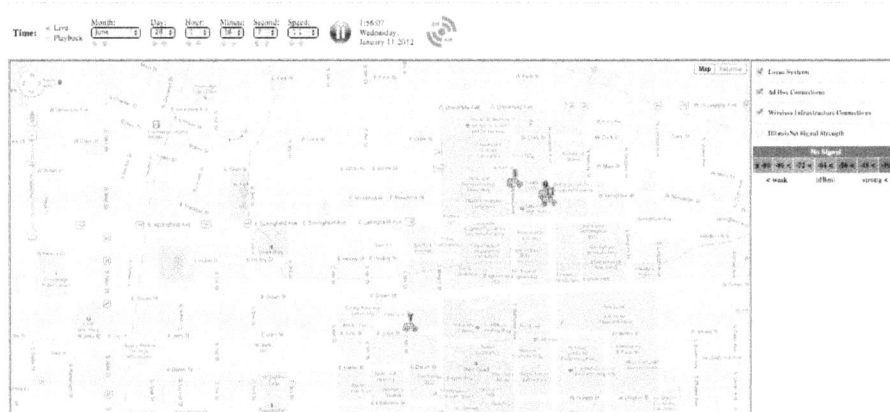

Figure 2: The IVT web interface

be implemented to identify parked vehicles and establish a connection when possible. Internet access is achieved by opportunistically associating to available networks. The Access Points are generally deployed to maximize the indoor coverage, and the signal that is leaked on the street is insufficient for providing a good connectivity. Our experiments on campus revealed that a moving vehicle, even when its speed is low, is rarely able to associate to one of these networks, and even when it does, the resulting bandwidth is very low and the duration of the connection short. To increase the amount of data that a vehicle can upload to the server during a contact, we implemented a multi-hop system. Vehicles that are parked near a building associate with an AP, and since they are static, they are able to establish a better and longer connection. On their 802.11g radio, these vehicles broadcast their presence, and act as gateways for moving vehicles. The ad-hoc connection is quicker (i.e., no association is needed), and has a better bandwidth, since both vehicles are on the street and there is no wall or other large obstacles. Using this connection the moving vehicle can upload large chunks of data to the parked vehicle.The parked vehicle can accept the upload from the moving car on a fast and reliable link, and then deliver the data to the internet at the slower rate supported by the link with the AP. Additionally, in many cases associating to an AP requires a significant overhead due to scanning on all channels looking for available APs, authentication, and obtaining a valid address. This overhead is unacceptable if the connection lasts only a few seconds, as is the case for a moving vehicle. Parked vehicles instead have a longer connection duration, thus the overhead represents a smaller fraction of the total exchanged data. Connection between moving vehicles and parked ones can be made quicker by using known channels and faster authentication procedures.

The most relevant projects related to the IVT and that inspired its design are DieselNet [1] which is designed for DTN communications using buses, taking advantage of fixed routes and predictable contacts, CarTel [3], in which a limited number of vehicles is equipped with radios to opportunistically connect to open APs, and BikeNet [2], which also uses opportunistic AP association to upload sensed data.

Our demo will be remotely connected to IVT, where some vehicles will be parked on the roadside, and some will be driving on campus. The running vehicles will collect data from the different sensors, upload them to the servers with a combination of direct wireless connection when possible, or using parked vehicles as gateways if they are avaiable. Visitors at the conference will be able to interact with our web interface shown in Figure 2 to see the live and historical data, and the notifications coming from moving vehicles when they pass by, and discover, vehicles that are parked on the road. We will show the different performance of direct connection to the available access points, versus the use of parked vehicles as gateways. At the demo site we will also have an example of the IVT hardware and we will demonstrate its functionalities and expandability.

[1] J. Burgess, B. Gallagher, D. Jensen, and B. N. Levine. MaxProp: Routing for Vehicle-Based Disruption-Tolerant Networks. In *IEEE INFOCOM 2006*, 2006.

[2] S. B. Eisenman, E. Miluzzo, N. D. Lane, R. a. Peterson, G.-S. Ahn, and a. T. Campbell. The BikeNet mobile sensing system for cyclist experience mapping. *SenSys '07*, 2007.

[3] B. Hull, V. Bychkovsky, Y. Zhang, K. Chen, M. Goraczko, A. Miu, E. Shih, H. Balakrishnan, and S. Madden. CarTel: a distributed mobile sensor computing system. *Conference On Embedded Networked Sensor Systems*, 2006.

[4] N. Thompson, R. Crepaldi, and R. Kravets. Locus: a location-based data overlay for disruption-tolerant networks. In *CHANTS '10*, 2010.

# Gossipmule: Scanning and Disseminating Information between Stations in Cooperative WLANs

Mónica Alejandra Lora Girón, Alexander Paulus, Jó Ágila Bitsch Link, Klaus Wehrle
Communication and Distributed Systems - Informatik 4
RWTH Aachen University
Aachen, Germany
{lora, paulus, jo, klaus}@comsys.rwth-aachen.de

## ABSTRACT

In Cooperative WLAN scenarios, the lack of a centralized management, the existence of many administrative domains and the current association process in wireless networks make it difficult to guarantee the quality that users expect from services and networks.

We present Gossipmule, an agent for wireless nodes that enhances the QoE perceived by users in Cooperative WLANs. Gossipmule uses mobile Crowdsensing between the wireless nodes to collect and disseminate information regarding the network. This information is used by the agent to have a more assertive association when making decisions regarding the user-AP association.

## Categories and Subject Descriptors

C.4 [**Computer Systems Organizations**]: Performance of Systems

## General Terms

Performance

## Keywords

Quality of Experience, Crowdsensing, Network-aware association, opportunistic networking, Cooperative WLANs

## 1. MOTIVATION

Nowadays connectivity is a must for users in many regions of the planet. This connectivity is defined by users in terms of an access to services anywhere, anytime and with any device. Consequently, according to Cisco [3], mobile data traffic will increase 26 times between 2010 and 2015 in a global scale as a consequence of the growing popularity of mobile devices. These trends are the engine of the development of strategies that take advantage of opportunistic communication.

One field where opportunistic communication could play an important role is in the enhancement of the Quality of Experience (QoE) perceived by wireless users. For instance, the current association services in IEEE 802.11 networks lack QoE mechanisms to enhance the connectivity experience of the user; a poor selection mechanism and high la-

tency are the main problems which are addressed by our proposal.

We propose to use Crowdsensing [4] as a mechanism to gather the required information in a decentralized way for an optimal association, which satisfies the quality requirements of users. In our approach, users collect relevant information regarding the current performance capabilities of the Access Points (APs) that are reachable for their user interfaces, as well as features related with the Basic Service Set (BSS) in which the user has been associated (i.e. realtime traffic patterns).

This sensing provides a partial knowledge about the network topology and conditions for the user. Afterwards we take advantage of the opportunistic communication between mobile users. The user exchanges his information with other users, in order to enhance his association decisions regarding the most suitable AP according with his necessities.

In this paper we address the research challenges using crowdsensing for the enhancement of the QoE perceived by the user. Based on this idea we propose a novel architecture using a sofware agent in the WLAN.

## 2. ASSOCIATION SERVICES IN IEEE 802.11

The three association related services in IEEE Std.802.11 [5] are association, reassociation and disassociation. Before a user can transmit information in a WLAN, the user or station (STA) should be first associated. During the association and reassociation, the STA scans the medium for Beacon frames to know what APs are present. After this, the STA chooses an AP in a BSS based on the strongest RSSI (Receive Signal Strength Indicator) value.

This process does not lead to the selection of the best Access Point (AP) that satisfies the connectivity requirements of the user, since it does not consider relevant metrics that have a strong impact on the performance [6] [1].

To improve the current association process and hence the QoE besides the RSSI value, additional information regarding the AP capabilities and performance is required by the user to make a smarter decision, i.e. number of associated users, load of the AP, airtime, among others.

The same problem affects the handoff process in WLAN. When the user is changing his association due to the availability of the AP or to the mobility, the user chooses an AP to establish a new association based on the RSSI value.

This QoE problem is exacerbated due to the high handoff latency that occurs during the scanning process of the handoff. Mishra et al. [6] have demonstrated that the reassociation delay in WLAN is about 50-400 ms and over of

90% of this time corresponds to scanning. It is important to point out that streaming, VoIP, among others applications are sensitive to high delays, thus latency handoff has a high impact on the QoE of the users. This effect is even more dramatic if we take into account that during the reassociation process the STAs are not able to receive or transmit information.

This reassociation latency would be reduced if the STAs have previous knowledge about the network topology and conditions, before the scanning process starts. Therefore, if we could reduce this reassociation latency, the effective transmission time will be increased, enhancing the perceived quality by the user.

Many approaches have been proposed in order to optimize the way that the STAs choose the AP [2] [1], but those approaches are centralized, demand more resources (additional interfaces) or depend on the information provided solely by the AP.

## 3. APPROACH IN DETAIL

### 3.1 Gossiper Design

The main part of this proposal is the gossiper agent which handles the collection, storage, evaluation and exchange of data. The mobility of wireless users is a feature that can be exploited in order to exchange information between STAs and collaborate in the enhancement of the overall QoE of the network.

When a user is scanning for a new AP, parameters such as load of an AP or the number of associated STAs will permit the user to take a more accurate decision when choosing a new AP to associate with. This information is broadcasted by the AP as beacon stuffing using some of the QoS capabilities that are defined in the IEEE 802.11e Standard. Additional metrics are calculated by the STA (i.e. airtime) based on the information that is sent by the neighboring STAs and by the AP.

Data exchange is done by using a part of the IEEE Std. 802.11v which supports multiple networks. Legacy 802.11 restricts data exchange to one network but as a new node does not have any association to an AP, the data has to be transmitted in a separate ad-hoc network. Therefore, using the IEEE Std. 802.11v, we are able to form a new network with a fixed SSID for our gossiper traffic. After the collection of this information, the user has to extract relevant information that allows him to build an AP preference list. In this way, when a STA requests data regarding the network topology, the agent disseminates opportunistically his current AP preference list.

Upon gathering a new data set for a potential list entry, the agent evaluates the data with regard to QoE and stores the AP in the list. Data which is not usable, for example if the data set is too old, is dropped and the list remains unchanged.

Additionally, based on the spatial and temporal relationships between wireless nodes. The STAs in our approach are able to disseminate information that already has been downloaded from other STAs. The sharing of data between the users increase the performance of the BSS reducing the overload of the AP.

Our approach is legacy compatible since STAs which do not support the gossiper module will then just ignore this additionally transmitted data.

### 3.2 Implementation

To get the gossiper module working as intended, some changes have to be made in the network device in addition to adding the gossiper agent as an application.

On the AP side we have to enable data gathering (e.g. number of STAs connected to the AP, load of the AP) and alter the beacon generation to fill in the additional data via beacon stuffing.

On STA side the gossiper module has to be implemented in the application layer. Additionally, we alter the MAC layer behavior to serve three additional purposes. First, extract QoS data from beacon frames and the gossiper data from probe frames, which we currently use as a wrapper. Second, add additional data provided by the agent to sent frames and third, enable virtual network support using specifications provided by 802.11v. The gossiper agent itself consists of an array containing the stored AP list including the data we evaluated to build it as well as management functionalities to handle requests and list exchange.

## 4. FUTURE WORK

Currently we are evaluating the presented approach using ns-3 as a network simulator in order to measurement the performance of opportunistic communication. Later on we would like to evaluate the consistency of our results in a testbed implementation.

Our future work will mainly focused on (1) Analyzing the Human mobility traces to make an smarter decision regarding target users with whom to exchange information, (2) Data quality with respect to the integrity and validity of the data that is exchanged and (3) Privacy issues during the opportunistic communication.

## 5. ACKNOWLEDGMENTS

This research was funded in part by NRW State within the B-IT Research School.

## 6. REFERENCES

[1] A. Balachandran, P. Bahl, and G. Voelker. Hot-spot congestion relief in public-area wireless networks. In *Mobile Computing Systems and Applications, 2002. Proceedings Fourth IEEE Workshop on*, pages 70 – 80, 2002.

[2] V. Brik, A. Mishra, and S. Banerjee. Eliminating handoff latencies in 802.11 wlans using multiple radios: applications, experience, and evaluation. In *Proceedings of the 5th ACM SIGCOMM conference on Internet Measurement*, IMC '05, pages 27–27, Berkeley, CA, USA, 2005. USENIX Association.

[3] Cisco. Cisco Visual Networking Index: Global mobile data traffic forecast update, 2012-2015. [Online] Available: http://www.cisco.com/en/US/solutions/collateral/ns341 /ns525/ns537/ns705/ns827/white$_p$aper$_c$11 − 520862.*html*.

[4] R. Ganti, F. Ye, and H. Lei. Mobile crowdsensing: current state and future challenges. *Communications Magazine, IEEE*, 49(11):32 –39, november 2011.

[5] IEEE. Std. for information technology - telecommunications and information exchange between systems - local and metropolitan area networks - specific requirements part 11: Wlan medium access control (mac) and physical layer (phy) specifications. *ISO/IEC 8802-11 IEEE Std 802.11 Second edition 2005-08-01 ISO/IEC 8802 11:2005(E) IEEE Std 802.11i-2003 Edition*, pages 1 –721, 2005.

[6] A. Mishra, M. Shin, and W. Arbaugh. An empirical analysis of the ieee 802.11 mac layer handoff process. *SIGCOMM Comput. Commun. Rev.*, 33:93–102, April 2003.

# Sensing Multi-dimensional Human Behavior in Opportunistic Networks

Sabrina Gaito
Università degli Studi di
Milano, Italy
gaito@dsi.unimi.it

Elena Pagani
Università di Milano, Italy
CNR-IIT, Italy
pagani@dico.unimi.it

Gian Paolo Rossi
Università degli Studi di
Milano, Italy
rossi@dico.unimi.it

Matteo Zignani
Università degli Studi di
Milano, Italy
matteo.zignani@unimi.it

## ABSTRACT

The massive spread of small personal devices equipped with different radio technologies is enabling the formation of a heterogeneous wireless networking platform on top of which new mobile computing services are deployed to flexibly and ubiquitously reach a target user. With the emerging of ubiquitous wireless communications, mobile applications are becoming highly personalized and influenced by user location, mobility, social attitudes and interests, or, shortly, by his/her behavior. In this work, we propose a framework having the aim of capturing and allowing the modeling the multiple dimensions of the human behavior. In order to measure and perform a cross analysis of those dimensions, we briefly describe an Android-based application able to collect face-to-face encounters and online social relations of a certain set of users.

## Categories and Subject Descriptors

I.6 [**Simulation and Modeling**]: Model Validation and Analysis

## Keywords

Opportunistic networks, mobility models, Android platform

## 1. INTRODUCTION

The massive spread of small personal devices equipped with different radio technologies is enabling the formation of a heterogeneous wireless networking platform on top of which new mobile computing services are deployed to flexibly and ubiquitously reach a target user. With the emerging of ubiquitous wireless communications, mobile applications are becoming highly personalized and influenced by user location, mobility, social attitudes and interests, or, shortly, by his/her behavior. It has been argued that mobility and sociality are tied one another and that both are essential to the design and the delivery of behavior-sensitive mobile services [3, 4, 5, 6]. Mobility is defined in the space-time plane, while sociality is defined in the social attitude-time plane. We

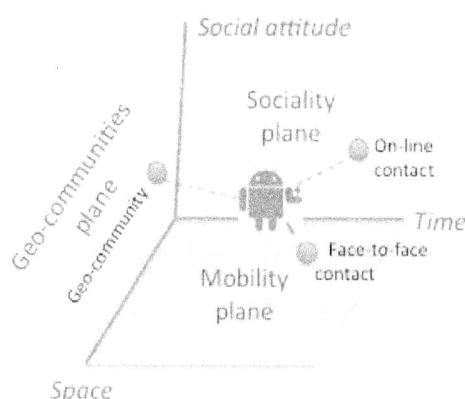

Figure 1: Three dimensions of human behavior.[1]

name 3-dimensionality of human behavior (for short H3D) the integration of space, sociality and time (Fig.1).

Our final goal is to build a unifying modelling framework in which the relationships among the three behavioral components naturally emerge and are well described to drive the design and the deployment of behaviour-sensitive services in an opportunistic networking scenario. To date, mobility and sociality planes have been considered by several studies but only separately. By contrast, the third plane, the social attitude-space plane, has been studied only in a few works, e.g. in [1] where geo-communities are defined as the combination of the spatial concept of location with the social concept of community. As far as we know, only the Stumbl project [3] has started combining mobility and sociality.

The main constraint to a wider combined analysis is related to the lack of rich datasets where H3D is represented. This poster has a twofold objective: the first is to describe the Android-based application we developed to generate such a dataset; the second is to utilize and extend the existing analysis methodology, based on the concept of geo-community, to obtain a unified modelling.

---

[1] Portions of this figure are modifications based on work created and shared by Google and used according to terms described in the Creative Commons 3.0 Attribution License.

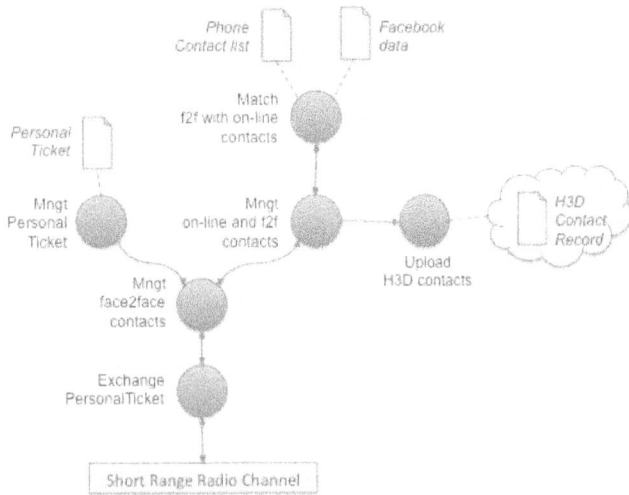

**Figure 2: Functional architecture of the Android-based application.**

## 2. SENSING THE THREE HUMAN DIMENSIONS

We designed an Android-based application, we name it *face2face*, that enables to detect and record the contacts in the real world, their location and enrich them with the contacts the involved people have in their on-line social experience. We focus on face-to-face (intentional) contacts because they tell more then extemporaneous contacts about people' sociality and because their expressiveness can be augmented by adding social contacts the encounters have on-line. To the purpose, when a face-to-face encounter is detected the application enriches data about the contact in the mobility plane with data about the sociality plane and concerning, for instance, friendship in a social network and phone numbers in the personal contact list. As far as we know, this approach represents a novelty in the area of similar experiences and enables to collect a very rich dataset.

In Fig.2, we show the functional architecture of the *face2face* application. The installed application is configured with the user's personal ticket that contains: an optional photo, the name, the phone number and the facebookID. Bluetooth is used to detect the contact, to exchange tickets, and to record the contact start and duration. To detect face-to-face contacts the ticket exchange is enabled by explicit authorization of the users via simple popup menu. The use of a popup menu also simplifies the BT criticalities of synchronizing devices and electing the piconet master station. When receiving the personal ticket of the encounter, a device seeks for possible matching in the phone contact list and facebook data, and records the matching results. The obtained contact descriptor is uploaded to the server.

## 3. ANALYZING THE THREE HUMAN DIMENSIONS

The analysis of H3D needs specific methodology and models able to operate in the three planes described in Fig.1. To operate in the geo-community plane, we developed a methodology to extract social relationships from GPS traces; the system (node, geo-community) is represented as a bipar-

tite graph whose projections on nodes indicate the strength of the relationships [2]. As regards the mobility plane, a new mobility model, named Geo-CoMM [7, 8], has been defined. People move within a set of geo-communities, following speed, pause time and choice rules whose distribution is obtained by the statistical analysis of large GPS datasets; similarly, inside a geo-community, people move according to a Levy walk. Geo-CoMM is able to properly reproduce not only the spatial and temporal features, but also the local and global social properties observed in real mobility datasets.

The analysis of H3D imposes to expand our tools to include in this general framework more tied social relationships as obtained by face-to-face and on-line contacts, and study their interactions with space-time patterns. To the purpose, we are using measures of similarity on graph to analyze the time-varying graphs deriving from different social interactions: nodes are the same, but communities derive from geo-community, on-line social network and face-to face contacts. This allows, for instance, to detect nodes with higher multi-dimensional centrality, on top of which to build algorithms for opportunistic routing.

## 4. ACKNOWLEDGMENT

The authors are grateful to Valerio Arnaboldi for the useful contributions during the implementation phase. This work was funded partially by the Italian Ministry for Instruction, University and Research under the PRIN PEOPLENET (2009BZM837) Project, and partially by the European Commission under the FP7 SCAMPI (258414) Project.

## 5. REFERENCES

[1] S. Gaito, G.P. Rossi, M. Zignani. From mobility data to social attitudes: a complex network approach. In *Proceedings of the NEMO Workshop*, 2011.

[2] S. Gaito, G.P. Rossi, M. Zignani. Extracting human mobility and social behavior from location-aware traces. In *Wireless Communications and Mobile Computing Journal*, Wiley and Sons ed., 2012.

[3] T. Hossmann, F. Legendre, G. Nomikos, T. Spyropoulos. Stumbl: Using Facebook to Collect Rich Datasets for Opportunistic Networking Research. In *Proc. Fifth IEEE AOC Workshop*, Lucca, Italy, June 2011.

[4] W.-J. Hsu, D. Dutta, A. Helmy. Profile-Cast: Behavior-Aware Mobile Networking. In *SIGMOBILE Mob. Comput. Commun. Rev.*, 12(1), 2008, 52-54.

[5] E. Pagani, G.P. Rossi. Reasoning about Multicast in Opportunistic Networks. In *Proc. 5th IEEE AOC Workshop*, June 2011.

[6] K.K. Rachuri, C. Mascolo, M. Musolesi, P.J. Rentfrow SociableSense: Exploring the Trade-offs of Adaptive Sampling and Computation Offloading for Social Sensing. In *Proc. 17th MobiCom*, 2011.

[7] M. Zignani. Human mobility model based on time-varying bipartite graph. In *Proc. 12th IEEE WoWMoM* – PhD Forum Session, 2011.

[8] M. Zignani. Geo-CoMM: a mobility model based on geo-communities. In *Proceedings of the 9th WONS*, 2012.

# From Ego Network to Social Network Models

## [Extended Abstract]

Marco Conti, Andrea Passarella, Fabio Pezzoni
CNR-IIT, via G. Moruzzi, 1 - 56124 Pisa, Italy
{m.conti, a.passarella, f.pezzoni}@iit.cnr.it

## ABSTRACT

The development of routing protocols for human pervasive networks can not disregard contextual information. Social network models can help researches giving information about human behaviors, e.g. estimating contact opportunities between nodes. In a previous work we defined a model for ego networks while in this work we present a method to integrate ego networks and obtain synthetic social networks ready-to-use for the development of routing protocols.

## Categories and Subject Descriptors

H.1.0 [**Models and Principles**]: General; C.2.1 [**Computer-Communication Network**]: Network Architecture and Design—*Store and forward networks*

## Keywords

Human Pervasive Networks, Social Networks Models

## 1. INTRODUCTION

In the last decade the proliferation of personal computing devices, such as smart-phones and PDAs, led researches to envisage a *human pervasive network*. Opportunistic connections among the devices would able end-users to share multimedia contents and to access to services without the support of an existing infrastructure. The development of routing protocols for opportunistic networks is an ongoing challenge because the topology is very unstable and dynamic. Moreover, only considering topological information to build network protocols is not enough to guarantee fast and effective circulation of data. There is the need to complement topological information with contextual information, and *human behaviors* must be considered as the key aspect. Information about human social relationships and interactions could help nodes to estimate information such as contact opportunities, contact frequencies and trust levels [1].

Our research activity exploits the results in the field of social anthropology in order to model human social networks, which are networks formed of the individuals connected to each other by social ties. In a previous work we defined a model for the generation of ego networks, i.e. social networks from the point of view of a single individual [2]. In this work we extend our model to integrate ego networks to obtain a social network formed of thousands of individuals.

Intuitively, to generate social networks instead of a set of disconnected ego networks, we need to define rules by which nodes in a given ego network can also be part of other ego networks, and with which role. Synthetic social networks, generated in simulation, show well-know properties discovered in real social networks.

## 2. EGO NETWORKS

An ego network is a simple form of social network defined from the point of view of an individual, composed of the individual (the "*ego*") and the set of social ties between the individual and other people ("*alters*"). Social ties can be classified into different type of relationship such as: kin, friends, neighbors, work colleagues, etc.. They can be strong (i.e. with very close relative and friends) of weak (i.e. with neighbors or work colleagues). The strength of a social tie can be influenced by several factors but the "*emotional closeness*" is considered to be the most influential [6].

The most remarkable results about the properties of ego networks were obtained by Dunbar et al. who demonstrated that social ties in an ego network are organized according to nested shells of increasing size and decreasing average strength [6]. The innermost shell includes the individuals having the strongest social ties with the ego and it has average size of 5. The outermost shell correspond to the whole ego network and anthropological study demonstrate that its size is around 150 individuals. The number 150, also known as "*Dunbar number*", represents the limit on the amount of social relationships an individual can maintain because each relationship consumes both cognitive and time resources and these resources are limited.

### 2.1 Ego Network Model

In [2] we define a generative an ego network model based on an iterative procedure which adds new social ties to a network until the resource budget of the ego is exhausted. For convenience we modeled the resource budget assigned to each ego with an amount of time, since cognitive resources are hard to quantify.

Our network model considers a three-level structure in which shells are called "*support clique*", "*sympathy group*" and "*active network*" with average size respectively 4.6, 14.3 and 132.5 [experimentally observed in literature][1]. Basing on experimental evidences by Dunbar, our model assumes a linear correlation between the sizes of the support clique and

---

[1]Dunbar described another shell called "*band*" with average size of 35 however, since there are not accurate information about its properties, we did not included it in our model.

the sympathy group shells. On the contrary, the model does not consider any correlation between the active network size and the sizes of the other shells [6]. As suggested in [6], the emotional closeness level is the key parameter to consider in order to select in which layer a social tie has to be included. Our model exploits the emotional closeness distribution derived by Dunbar [6], to define the emotional closeness level for each tie within a given shell. We also defined a function $h$ that binds the emotional closeness level ($e$) and the amount of time invested in a social tie ($t$).

The algorithm initializes each ego with a budget of time and with the sizes of the support clique and sympathy group shells. We use a distribution of the time budget with mean value equal to 1720 hours, i.e. the 20% of the time in a year. For the size distributions of the support clique and the sympathy group, we exploited the results in literature. After the initialization, the algorithm starts creating new social ties for the support clique shell until it reaches the target size, subsequently it does the same for the sympathy group. For the outermost shell, the algorithm adds new social ties until the budget of time is totally exhausted [2].

## 3. SOCIAL NETWORKS

We experimentally demonstrated that our model of ego network is able to generate synthetic ego networks those satisfy well-know properties of the real ego networks [2]. However, in order to use our model for the development of human pervasive networks, we need social networks consisting of several thousands of nodes/individuals. As a social network can be viewed as the sum of ego networks, in this work we present a method for the integration of ego networks.

This method is based on the social network properties. Social networks have a high value of the clustering coefficient because of the emergence of communities, while the average path between two nodes/individuals is short. This phenomenon is widely know as *small world property*. For example, for the Facebook network (721 million of nodes), the clustering coefficient is $0.14^2$ and the average distance between pairs of nodes is 4.7 [7]. Additionally, social networks' structure is influenced by geographic constraints because it is more likely to form a tie between geographically close nodes than between distant nodes [5].

### 3.1 Ego Network Integration

Our strategy for the integration of the ego networks is to consider each node of a social network as an ego. A social network is initialized with $N$ nodes, than for each node it is executed the procedure for the ego network formation, described in Section 2.1, in a parallel fashion, adding social ties between nodes belonging to the network. In order to ensure the desired network properties we had to define a method that, given an ego, selects the proper destination node for each social tie to add. To this end we considered the mechanism of *triadic clousure*, introduced by Granovetter [3]: when the algorithm has to form a new tie, it tries to close a triangle, that is it selects a node 2-hops distant as tie's destination, preferring to follow links with a high level of emotional closeness. Triadic closure ensures a high level of clusterization but, in order to ensure the small world property we have also to consider the possibility of create

---

[2]In large random networks the clustering coefficient tends to zero, therefore 0.14 is commonly considered as remarkable.

Table 1: Measures of a Generated Network

| nodes | 50.000 |
|---|---|
| avg degree | 134.1 |
| avg path length | 2.84 |
| cluster coefficient | 0.134 |

a tie with a node randomly chosen. This mechanism allows the average path length to be small. In our model we introduced a parameter $p$ that is the probability to use the triatic clousure mechanism rather than select the target node randomly. The random node selection has to take into account the geographical constraints, therefore we introduced geographical information placing the nodes in a one-dimensional space. Basing on the results in [5], the probability to create a tie with a node, decreases with the distance ($d$) from it, in compliance with a power-law probability distribution. The density function of this probability distribution is

$$f(d) \propto \begin{cases} d_{min}^{-\alpha} & \text{for } d < d_{min} \\ d^{-\alpha} & \text{otherwise} \end{cases} \qquad (1)$$

where $\alpha$ is the exponent of the power-law function and $d_{min}$ is a threshold under which $f$ is constant.

## 4. RESULTS

Preliminary results of our experiments are very promising. In Table 1 we present some measures of a network of 50.000 individuals, generated setting $p = 0.6$, $\alpha = 1.5$ and $d_{min} = 0.002$. Measures in the table and other properties, like the correlation between the degree and the cluster coefficient and also the cluster coefficient distribution, are similar to those ones presented in literature for real social network [4, 7].

## 5. ACKNOWLEDGMENTS

This work was partially funded by the European Commission under the SCAMPI (FP7-FIRE 258414), RECOGNITION (FP7 FET-AWARENESS 257756), and EINS (FP7-FIRE 288021) projects.

## 6. REFERENCES

[1] M. Conti et al. From opportunistic networks to opportunistic computing. *Communications Magazine, IEEE*, 48(9):126–139, 2010.

[2] M. Conti et al. A model for the generation of social network graphs. In *WoWMoM, 2011 IEEE International Symposium on a*, june 2011.

[3] M. Granovetter. The strength of weak ties. *American journal of sociology*, pages 1360–1380, 1973.

[4] A. Mislove et al. Measurement and analysis of online social networks. In *Proceedings of the 7th ACM SIGCOMM*, pages 29–42. ACM, 2007.

[5] J. Onnela et al. Geographic constraints on social network groups. *PLoS one*, 6(4):e16939, 2011.

[6] S. Roberts. Constraints on social networks. In *Proceedings of the British Academy*, volume 158, pages 117–138, 2010.

[7] J. Ugander et al. The Anatomy of the Facebook Social Graph. *ArXiv e-prints*, Nov. 2011.

# Twitter in Disaster Mode:

## Smart Probing for Opportunistic Peers

Theus Hossmann, Dominik Schatzmann, Paolo Carta, Franck Legendre
Communication Systems Group
ETH Zurich, Switzerland
firstname.lastname@tik.ee.ethz.ch

## ABSTRACT

Recent natural disasters (earthquakes, floods, etc.) have shown that people heavily rely on platforms like Twitter to communicate and organize in emergencies. To mitigate communication outages due to broken infrastructure during such events, it was proposed to make mobile Apps "disaster ready": In normal operation mode, they use cellular infrastructure to communicate, whereas in disaster mode, they rely on opportunistic communication and disseminate messages in a peer-to-peer manner, using Bluetooth, WiFi ad hoc or WiFi Direct. Such *hybrid* applications can use the infrastructure to prepare for a network outage.

In this poster, we present preliminary work towards such a hybrid solution to solve the ad hoc connectivity problem. For security reasons, discoverability of devices (e.g., in Bluetooth) is often restricted to short time periods – a limitation which prevents us from building usable opportunistic networks for disaster relief. However, if we are able to predict which peers are within transmission range, we can circumvent this limitation: Instead of scanning for discoverable devices, we can probe for potential peers and simply connect if we find them in range. In our approach, the disaster ready App obtains a list of predicted potential peers from a central server during normal operation mode, which is then used for probing in disaster mode. We discuss the challenges and limitations of such an approach and sketch how to predict contacts based on location and/or social information.

## Categories and Subject Descriptors

C.2.1 [**Network Architecture and Design**]: Wireless Communications

## Keywords

Opportunistic networks, Smart Probing

## 1. MOTIVATION

As recent natural disasters have shown, many affected people turn to online social networks (OSNs) such as Facebook and Twitter to communicate, organize and share updates with their friends and other victims in emergencies [1, 2]. However, today's OSN applications require Internet connectivity and are therefor vulnerable to breakdowns of the communication infrastructure (3G, WLAN and fixed lines). Thus, it is likely that at the time when the victims of a disaster need their favorite OSN the most, it does not work. For example, an analysis has shown that in the aftermath of the

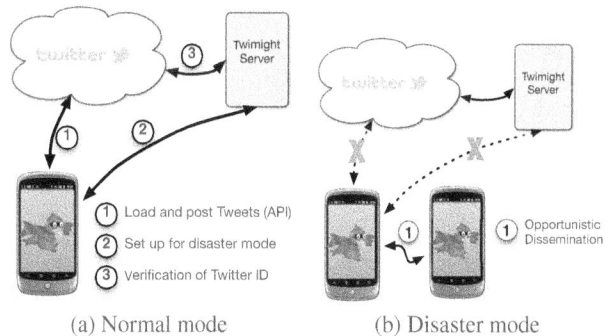

(a) Normal mode      (b) Disaster mode

Figure 1: Twimight operation modes.

earthquake and tsunami in Japan in 2011, while the backbone of the network remained largely unaffected, some sites were disconnected from several hours to days [3].

To mitigate this problem and make OSNs more useful for disaster relief, it was proposed to employ opportunistic networks: If no infrastructure network is operational, messages are spread directly among mobile devices (e.g., smart phones) using wireless peer to peer technologies such as Bluetooth, WiFi Ad Hoc or WiFi Direct. To this end, we have implemented *Twimight* [4, 5], a Twitter application for Android phones which is equipped with a "disaster mode". In normal operation mode, Twimight uses the Twitter API [1] to offer normal Twitter functionality. If connectivity is disrupted, the user can switch to the disaster operation where Tweets are spread from device to device via Bluetooth. The two operation modes are visualized in Figure 1. In such a *hybrid* (two mode) application, the connected operation can be used to prepare for disconnected phases. To support the disaster mode and set up everything for opportunistic operation (e.g., distribute digital certificates to sign and encrypt messages [5]), Twimight uses the *Twimight Server*.

While the centralized server support eases many problems that are typically very hard in completely distributed applications (see [4] for a list of challenges), many challenges remain. A fundamental one we focus on here is that of connectivity: Today's mobile operating systems provide only very limited support for wireless peer to peer technologies. Table 1 summarizes the situation for the two most widespread operating systems, Android and iOS[2]. To overcome this limitation and build opportunistic networks in natural

---

[1] http://dev.twitter.com

[2] This describes the situation as of Android 4.0, early 2012. For iOS, the situation is even worse since there is limited support for background services – a requirement for scanning for opportunistic peers. Currently, there is no evidence that this situation will change in the near future.

| | Android | iOS |
|---|---|---|
| **Bluetooth** | Discoverability requires user interaction and is limited to 5 min | Officially supported for gaming only |
| **WiFi Ad Hoc** | Not supported | Can connect only to existing ad hoc network (not create) |
| **WiFi Direct** | Discoverability requires user interaction and pairing with PIN [6] | Not supported yet |

Table 1: Limitations in mobile operating systems.

disasters (where we assume that security requirements are not as stringent as in normal operation), users must root their phones – an infeasible solution for applications aiming at ease of use and wide adoption. Here, we propose a solution to avoid the scanning (and hence the limitation due to restricted discoverability) by probing for devices which have high probability of being in transmission range. We focus on Bluetooth because of its wide spread, but we believe that similar solutions could be feasible for WiFi Direct as well in future.

## 2. SMART PROBING FOR PEERS

Even without a previous scan for Bluetooth peers, the Android API[3] allows to open an insecure RFCOMM connection to a given Bluetooth MAC address.[4] Prerequisites for a solution without scanning is that the devices know the MAC addresses of potential peers and – to avoid wasting energy and time, probing for unpromising peers – which of them are to be expected in transmission range. Thus, the question is: *How can a mobile device know about potential peers to probe for?*

In the *hybrid* architecture of Twimight we use the Twimight server for this purpose. At regular intervals (e.g. every 24h), our Twimight application sends the user's main locations along with its MAC address to our server. In return, our server provides a list of potential peers based on a similarity metric predicting potential peers to probe for. The request is a JSON formatted HTTP POST containing the OAuth tokens for authenticating the Twitter user, the Bluetooth MAC address for identifying the device, as well as a list of timestamped waypoints – location samples taken every 6h from the phone's GPS and/or other localization means. The server replies to this request with a JSON encoded array of the 10 MAC address with the highest similarity scores. For now we simply rely on a user location profile based on geo-localization samples. In future work, we plan to include other profile information such as Twitter followers, checkins and any users' online social information in general.

## 3. DISCUSSION AND OUTLOOK

In the following we first put our approach in perspective with other ad hoc communication protocols besides Bluetooth. We discuss then the security and privacy issues raised by our approach and plans for future work.

The WiFi Direct standard was not designed with opportunistic communication in mind. It mostly targets home networking were devices (e.g. smartphone, printer, game console) pair to each other using PIN codes. For now, WiFi Direct requires users manually

initiating the discovery of devices and then pair using a PIN code. Our approach could help to exchange PIN codes between nearby peers instead of MAC addresses. Another standard, Bluetooth 3.0 + HS [7], proposes to use Bluetooth to discover peers and then 802.11 for data transfers. It is unlikely that this standard will prevail against WiFi Direct. To bypass all these limitations and the lack of support of WiFi Ad Hoc, Trifunovic et al. propose WLAN-OPP [8], a more open standard than WiFi Direct specifically targeted at opportunistic networking with energy saving strategies. WLAN-OPP uses open 802.11 connection and delegates security to upper layers. This comes at the risk of layer 2 attacks but favors ease of discoverability and connectivity.

While we assume no security or privacy threats during disasters, our approach can still be misused for tracking one's MAC address in normal mode of operation. A malicious user could request potential MAC addresses at given locations and track one's whereabouts. We have to mention that with Twimight, a peer's Twimight or Twitter ID is not disclosed when opportunistically exchanging tweets. In order to map a MAC address to a real or online identity, an attacker would have to use other means. Nevertheless, our TDS server does not accept requests with single locations. Besides, we will limit the rate at which users can send requests and only allow for slight change in profiles between successive requests. To avoid Sybil attacks, we will request user interactions from time to time to cross-check the location samples.

Eventually, we intend to investigate the effect of parameters such as location sampling frequency and peer list size with the analysis of location traces and experiments in the future. The study of the performance-privacy tradeoff will be evaluated. We also plan to release a generic API which can be used also by other "disaster ready" applications or opportunistic applications in general.

## Acknowledgment

This work was partially funded by the European Commission under the SCAMPI (FP7 – 258414) FIRE Project.

## 4. REFERENCES

[1] A. Chowdhury, "Global pulse," http://blog.twitter.com/2011/06/global-pulse.html, 2011.

[2] A. Bruns, "Tracking crises on twitter: Analysing #qldfloods and #eqnz." Presented at the Emergency Media and Public Affairs Conference, Canberra, 2011.

[3] K. Fukuda, M. Aoki, S. Abe, Y. Ji, M. Koibuchi, M. Nakamura, S. Yamada, and S. Urushidani, "Impact of Tohoku earthquake on R&E network in Japan," in *Proceedings of the Special Workshop on Internet and Disasters*, 2011.

[4] T. Hossmann, F. Legendre, P. Carta, P. Gunningberg, and C. Rohner, "Twitter in disaster mode: Opportunistic communication and distribution of sensor data in emergencies," in *ExtremeCom*, 2011.

[5] ——, "Twitter in Disaster Mode: Security Architecture," in *Proceedings of the Special Workshop on Internet and Disasters*, 2011.

[6] WiFi Alliance, "Wi-Fi Peer-to-Peer (P2P) Technical Specification," March 2010.

[7] Bluetooth SIG, "Bluetooth Core Specification Version 3.0 High Speed (HS)," http://www.bluetooth.com/Bluetooth/ Products/Bluetooth_High_Speed_Technology.htm.

[8] S. Trifunovic, B. Distl, D. Schatzmann, and F. Legendre, "WiFi-OPP: Ad-Hoc-less Opportunistic Networking," in *ACM CHANTS*, 2011.

---

[3]API Level 10 and higher (Android 2.3.3).

[4]Here we focus on Bluetooth only and the applicability of our approach to WiFi Direct is discussed in Section 3.

# Author Index

www.ingramcontent.com/pod-product-compliance
Lightning Source LLC
Chambersburg PA
CBHW082107210326
41599CB00033B/6615